Other books by the Author (Amazon/B&N)

My 50 Years in High-Technology

No BS…Your MBA Primer

Mission Success in Space

The $Panic of 2020 I

The $Panic of 2020 II

The $Panic of 2020 III

The Decades

The Despots

Table of Contents

Book Dedication

To my NASA GSFC Flight Assurance Management and Systems Engineering Teammates and Colleagues whose work focus, dedication and individual contributions are too numerous to mention… I thank you all for a hell of a ride from Earth to Space and beyond!

CVOB Summer 2019

PREFACE

The purpose of this book is to document, in one place, the historical aspects of the many NASA GSFC missions that were performed during my tenure there. And to cover requirements that involved standard practices in designing, building, testing and launching spacecraft. In addition, the book describes the application of a new formal program of what we call Mission Success, a term so widely used at NASA, but vaguely defined.

In this book, we take a close look at this NASA mantra and endorse it as formal Standard Operating Practice (SOP) on all USA Space Program missions. What results is a higher confidence level of success and lower risk of failure. And, that effectively reduces the non-conformance cost of quality over many of the past spacecraft design, build and launch practices.

The book also covers several new Quality Management Systems tools and techniques such as, Barrier Analysis (BA). This rapid assessment tool is introduced so that it will provide an extra measure of Quality, Safety and Reliability.

Combined, all these tools & techniques will fill-in the many anomaly escapes and prevention gaps that exist in the present Assurance Management structure.

Finally, the book provides Management with on-going Continuous Improvement measures that need implementation for Mission Success in present and future spacecraft endeavors.

Chapter 1: Where this Mission Success (MS) Book Began! NASA GSFC!

As with all things man invents and needs to write books about, it always begins with a story or flash in one's mind that takes us all to new places or subjects that needs more exploring.

Well, let's begin there, perhaps it began with a brainstorming session where three long-time USA Aerospace colleagues (with nearly 100 total years' experience, including the moon landing, NASA Apollo Program) asked a question?

Just what is this thing we always here about at NASA, called Mission Success?

From day one at NASA, we were exposed to a new culture where all engineering teams, government and contractor employees, perhaps for the first time, all worked in harmony to accomplish high-risk space explorations with little thought of failure and maximum focus on success.

And always in the background was the necessity of concerning the welfare of our precious human resources…the USA Astronaut Corps. First and second generation, "***Right Stuff***" human beings who flew the many dangerous missions to outer space. Some were retired and we had the next generation to support the US National Space Transportation System or STS for short!

The three engineers were meeting to make a formal PowerPoint presentation to NASA Senior Management on a new long-term NASA Mission. If successful, we all would join the new Project to implement our findings in embedding front-end mission success requirements that became the foundation of the effort. To us, our PowerPoint presentation was the very first time a mission success pitch to these people was made on the subject. We all promised that if we got the time someday, we would write a book on our findings since it was a breakthrough on a subject everybody talked about but there was no hard writeups on the amazing subject.

There is not one aerospace manager or engineer who is serious about achieving their goals or objectives, who would pay any price for how to do it successfully. At NASA, it appeared to the public that this was accomplished like a walk in the park on a sunny Sunday afternoon. Take it from one who has been there, it wasn't!

So, what did we present and how did we sell it to management? I drafted the first slide show, defining what mission success was and how it could be implemented. Let's start there as shown below.

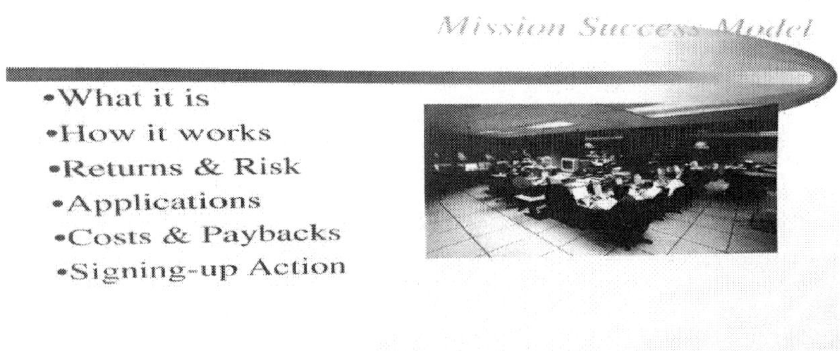

Exhibit MS-1 Mission Success Model Introduction Concept Slide

Exhibit MS-1 was designed to catch audience interest, as in reactions of "show me more", and we did, as the next slide shows.

We followed with the "how" message by listing more details in Exhibit MS-2. It showed, how we saw MS working for a NASA Project. We addressed typical Aerospace practices used by NASA Programs in the past, even the latest techniques such as: Risk Mitigation and Lessons Learned methods. However, we combined them into a new innovative roadmap to reveal our methods of a formalized mission success doctrine.

We did this by adding in new technological advances, tools and techniques and new US Aerospace Industry Standards such as AS9000. These requirements were embedded into hardware and software from initial design to fabrication to launch and even post launch stages.

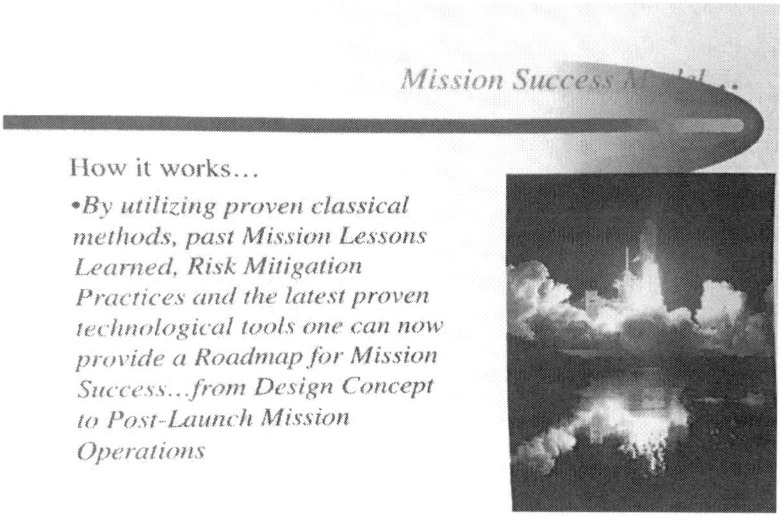

Exhibit MS-2: New Mission Success for Space Recipe

We got their attention at this point. You must remember our NASA audience is bombarded daily with ideas to the point that actual boredom sets in unless you have something new to sell…and we certainly did have! AIDA is always a good technique to use in making an effective presentation. This was the technique we used to close our deal!

Action (get audience attention) followed Interest (whet their appetite for more of what you are trying to sell) Desire (Get feedback ask questions, keep their interest!) Action (Close, close, close the deal!).

From here we went to specifics, next slide please. Exhibit MS-3

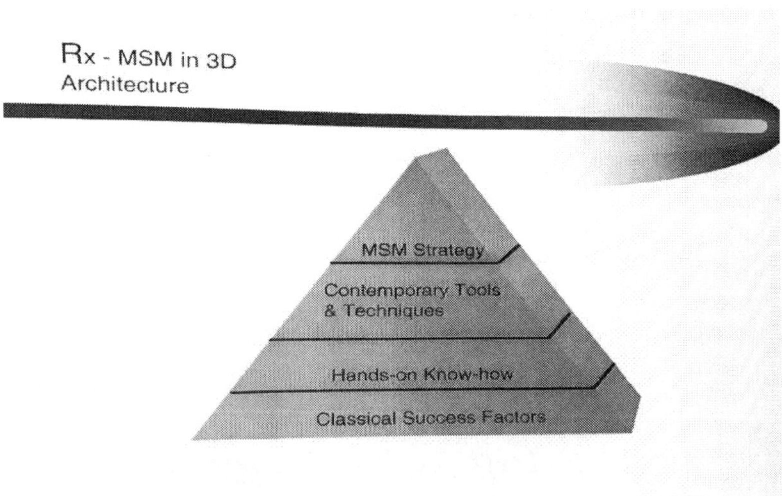

Exhibit MS-3: MSM in 3D Architecture

Exhibit MS-3 is seen as an engineering prescription that begins with a strategy (Your goal and how it is achieved). Followed by what we described as contemporary tools and techniques (CAD/CAM & Computer Modeling), hands-on knowhow (trained/certified personnel) and our classical Success Factors (later Chapter).

This was followed by a flowchart to show what it took to lay out a MS Plan and what elements it should contain. Guess what, we found that no Mission Success Plan existed. That got management's attention. Managers all fear one thing in business or government…failure! And as everybody knows there is always the scapegoat resulting…and nobody wants that. About 10 minutes into our discussion they knew we were serious…all eyes were on us…who the hell were we, I'm sure they were thinking…what magic wand (a DC thing!) did we possess. I'll just call it experience…and lots of it, hands on!

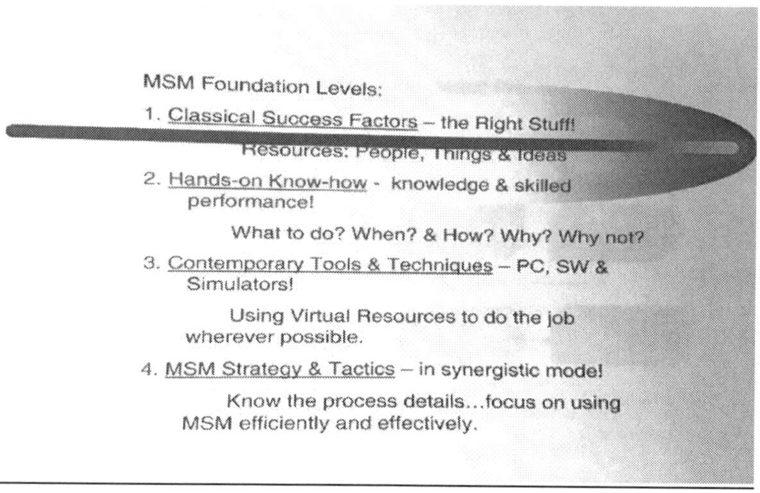

Exhibit MS-4 MSM Foundation Levels

Next, we showed Exhibit MS-4. We used the analogy of building a new home…starting with the foundation and building upwards to the support walls and roof. Each step combined the level plans all program managers deal with people, things and ideas, all in synergy to accomplish a goal successfully. But still not clear is how to put them all together?

That was where we came in…the fixers! Using our expertise, we could provide that extra insurance against risk, but not uncertainty as is often dealt with in a high-risk business.

Just before we finished our pitch the top manager said, "we have seen enough, you can come on board", which we did immediately and began our mission success journey.

This book will go into detail on the subject matter. I will cover all the details and "Success Factors" that were used for years at NASA and all its contractors.

In retrospect, as I review the term Mission Success, I cannot help to relate to what I call "Business Success" also! For the very ingredients to attain repeated business profitability is covered in most Management graduate and undergraduate training. Subjects like Management Planning (Strategy/Tactics), Staffing (Selection, Training, Supervision, Motivation), Organizing (by Program, Project or Classical Line and Staff functional teams), Directing (making/meeting goals/objectives, responsibility, authority and results), Controlling (delegating & accountability), Reviewing (PERT/CPM, Master Schedule progress, PDR/CDR), Budgeting (Variance Analysis, Plans Vs Actuals, Contingencies).

We will follow this flow in handling how to implement mission success to the high-risk business of designing, building, testing of space industry products. As the Table of Contents book outline shows, our journey will take us, using a management and systems approach, to cover such a complex subject as "***Mission Success in Space".***

Chapter 2. MS History I – The Dark Ages of Space Innovation

Before NASA was created in 1958, the USA was not sleeping, on March 3, 1915, NACA (National Advisory Committee on Aeronautics) was established to focus on the new science called aeronautics. Four key Laboratories were the outcome of that era and became essential resources in the later years. And they still are in existence today. First there was Langley, in Hampton VA. next, AMES at Moffett Field, CA. then Lewis Research, Cleveland (now Glenn) followed and Edwards Air Force Base, CA. The big accomplishment of NACA was having Chuck Yeager who piloted a Bell X-1 break the sound barrier on October 14, 1947 dropping from the bombay of a B-29.

The USA Space Age all began with Robert H. Goddard, Worchester, MA on March 16, 1926. He and his team launched 34 rockets from 1926 to 1942. He is considered the Father of Modern Rocketry. He published ***"A Method of Reaching Extreme Altitudes"***, in 1919, a classic influential work! It is in his memory that the USA NASA Goddard Space Flight Center (GSFC) was so named on May 1, 1959.

In Germany, a fellow named Dr. Verner von Braun (wrote 33 books according to Goodreads) was developing what we now know as the V1 & V2 Rockets at a place called Peenemunde, Germany, their Cape Canaveral in its days up until 1945. After the war, He and his scientist team were captured by the US Army (See Operation Paperclip) and taken to the USA, eventually settling in a small Southern town called Huntsville, Al. Now called the ***Rocket City*** because of the unique space history developed there. A great USA Space Museum now exists there and is worth a visit by all Americans, young and old!

Up to that time, late 1940s, Space rocket launches were to say the least, hit or miss. Many accidents occurred with this new technology. All was very quiet until our adversaries, the Russians, shocked the planet by launching their famous satellite called the **Sputnik** on Friday, October 4, 1957. This was followed by a **first** man in space flight when Yuri Gagarin on April 12, 1961 on Vostok I for one earth orbit! USA followed, less than one month later, when in May 5, 1961 with a sub-orbital flight by Alan B. Shepard, USN on a Redstone Rocket.

That led to more serious matters as far as space was concerned called Manned Space Flight! And a new post WWII leadership made sure of that with a USA President and Harvard graduate, WWII NAVY hero named John F. Kennedy. Not to take away from the last great President, Dwight D. Eisenhower, 5 Star General and WWII Supreme Commander, who first funded the launch of a new era called the Space Age!

The government funded the research/development and industry responded. One standout at the time was "The Martin Company" founded by Glenn L. Martin started in 1912 and ended in 1961 merging with Marietta, then in 1995, with Lockheed to be called Lockheed Martin today. Martin in the early days even taught William Boeing how to fly. Later in the 1950s, it became a missile producer with it's Vanguard and Titan missiles used in many early space missions, Explorer I, our first JPL designed USA satellite payload, being just one of them.

When I worked at Martin Marietta, Baltimore in 1983, Titan II was still an active program. I recall the "Good Old Boy" Program Manager who was still there from the early 1960s. It was the largest ICBM ever deployed by the USAF.

In the 1990s, the consolidation of the Aerospace business left only Boeing, Lockheed Martin, Northrup-Grumman/TRW and Raytheon (Hughes/GD). Gone are North American, McDonnell Douglas, Republic Aviation, GE, Ford, IBM, Vought/LTV, Fairchild, Collins Radio and more! This caused severe losses in innovation and competition in a critical high technology industry. The fact that only Boeing makes large commercial jets such as the 737MAX as a USA big league producer, should be a cause for concern for USA continued leadership in aerospace and space technology and all of USA citizens.

A visit to Kennedy Space Center (KSC) today, takes the Florida Visitor's Center tourist on missions from the 1950s to present day. In the Space Museum Park, one can see actual missiles of yesteryear from Bomarc to Lacrosse, even my good old US Army Nike Ajax and Hercules. The second tour takes the visitor to modern day Space Era missiles with even a mockup of the Space Shuttle for one to see. In addition, a nearby Space School adds to the resources available for the serious student. During their times the missile displays were state of the art in USA Space Technology, but we never thought of riding on them.

Chapter 3. MS History II – Manned Space Flight

The USA leadership was not completely surprised by the launch of the Sputnik! But, in our democracy, political infighting in DC and among the military services was over who would take the lead in this new era and obtain the necessary funding? That was what it was all about.

Russia first changed the infighting quickly. Soon after Sputnik, the USA spun off US Army Ballistic Missile Agency (ABMA) in Huntsville, AL to become NASA, a new civilian agency who was to take the ball and reach for the stars.

Up to this time, we had an arsenal of all the military services missile technology, all centered around Huntsville. I was amazed on my first trip there in 1959, to see the stadium-like display, all lite up and standing tall, the complete USA missile family! What a site to behold, as we came down the nearby hills of Mont Sano, into the city of Huntsville, AL.

Exhibit MS-5: US Army Missile Display in Huntsville Circa 1950s.

The Space Age in the USA, for our modern-day era, began in the 1950s. Once NASA was formed in 1958 by then President Eisenhower, for peacetime purposes, public interest became fascinated by the new venture. TV also was becoming a household entertainment source and the Space race was all the rage! So, we begin by refreshing our space journey. First, with the selection of our new space heroes the Mercury 7 Astronauts, then to Gemini and finally Apollo and the Space Shuttle Program!

First USA Space Missions: Project Mercury (1958-1963) – 6 space flights, Redstone and Atlas missiles.

Description: First man flight with Mercury Seven Astronauts chosen for first USA Space Flights.

Purpose: To test man's survivable Launch, Pre-orbit, Orbit and Post-orbit capabilities

Results: Mission Success. First flight overall, better than expected. Kudos: Allen B. Shepard (first USA in Space), John Glenn (first USA to circle the Earth), Scott Carpenter, Gus Grissom, Deke Slayton, Gordon Cooper, Wally Schirra.

Second Mercury Capsule lost on recovery phase due to escape hatch door opening prematurely, however, Astronaut Gus Grissom survived.

Lessons Learned: Training a valuable risk mitigator!

Program achieved goals resulting in Mission Success.

Second USA Space Missions: Gemini Program (1962-1966) – 12 flights, Titan II missiles

Description: Two-man crew space capsule for earth orbit.

Purpose: To test long duration missions, Rendezvous/Dock, perfect re-entry/landing understand Space effects.

Results: Astronaut Ed White performed first USA spacewalk, longest space flight, first rendezvous, first auto-re-entry to Earth.

Gemini 9 unable to perform space dock with Agena rocket stage. Gemini 12 with three space walks solved overheating from previous flights.

Lessons Learned: In Space work problems can be overcome, long space walks problems now overcome. Paved the way for moon mission.

Program achieved goals resulting in overall NASA Mission Success.

The following Exhibits MS – 6 & 7, are photos of those successful missions. The first shows Gemini 7 in space, on-orbit taken by another Gemini 6 mission also on orbit!

This picture of the Gemini 7 spacecraft was taken by the Gemini

Exhibit MS-6: Gemini 7 in Space photo taken by Gemini 6 crew

The second photo shows the first USA Spacewalk performed by Astronaut Ed White who has no fear of heights or doing the first space walk around. Hats off to this fearless USA Astronaut!

Exhibit MS-7: Astronaut Ed White performing first Gemini Space walk

Chapter 4: MS History III – Apollo to Space Transportation System (Shuttle)

Third USA Space Missions: Apollo Program (1963-1972) – 12 flights all with new Saturn launch vehicles.

Description: Three-man, three-part spacecraft (Command Module, CM, Servicing Module, SM & Lunar module, LM) designed for longer duration space exploration. The Lunar Excursion Module (LM) was now part of moon mission. And a new Launch vehicle called Saturn was tested for the longer duration moon mission (See Exhibit MS- 8).

Purpose: To achieve the USA goal of landing on the moon, in this 1960 decade, and safely returning to Earth.

Results: Neil Armstrong took control of the LM after the computer went into overload and safely landed the space craft. After both he (first man) and Buzz Aldrin (second man) walked on the moon and they safely returned to Earth as planned with Michael Collins, Command Module Pilot.

One mission Apollo 13 was made into a movie showing a close call incident where American ingenuity again and a determined Mission Control Command Director, Gene Krantz, that **"Failure Was not An Option"** brought the crew home safely. Six Mission achieved the USA goal of JFK made several years earlier.

Lessons Learned: The Mercury 7 Astronauts were right; it does take man to achieve space goals by applying human intervention to overcome failure situations.

Program achieved all its goals resulting in NASA Mission Success.

Exhibit MS-8: Photo of Saturn Rocket at KSC liftoff.

Let's go back to the beginning of the Space Age…the 1950s! As we tried over and over to launch successfully a missile to Space, we failed again and again. The Martin Company (later Martin Marietta, now Lockheed Martin) investigated and analyzed launch failures. At first, no one reason stood out amid many trials. Failure causes seemed to be everywhere as there are with all new inventions. However, a new group at Martin, making the Matador Missile, called Quality Control, after repeated failures, convinced management to institute a new part screening procedure. They discovered that once the first incoming inspection/testing of parts began that many parts procured were not meeting their published specifications. In fact, acceptable parts were the exception. And, once tested to extreme temperatures and shock/vibration requirements, part yields fell far short of quantities required to build a missile.

Months passed until a full requirements inventory was made available. Design engineers did their job, however, part reliability at the time was still theoretical! Reliability Engineering was in its infancy stage also and did not help matters. It too proved to be, in the long run, a missing success factor and function, in the magic recipe for achieving space flight mission success.

In addition, launch personnel and facilities also were coming down the learning curve adapting their new skills and continuously improving launch operations, especially at Cape Canaveral (now Kennedy Space Center KSC) in Titusville, Florida.

October 4, 1957 newspaper headlines around the world announced that Russia was first in space, with a small satellite called Sputnik circling the globe! Most Americans were shocked by this new technology! President Eisenhower then ordered our Army Ballistic Missile Agency, ABMA, (predecessor to NASA MSFC) in Huntsville, Al, and JPL (payload designer) in Pasadena, CA to get to work. The Dr. von Braun rocket scientist and his team of rocket experts at the WWII German Launch Site at Peenemunde, Germany, now in the USA, responded immediately.

Yes, there were a few mishaps between these times, Army, Air Force and Navy rivalries and DC politicians fighting for position. But I give special credit to LTG 3 Star General

Medaris, at Redstone, for his intestinal fortitude that led to our eventual success, at that time, in Space.

On January 31, 1958, using a souped-up Redstone, called Vanguard, Rocket City, USA helped launched the USA's first Satellite, a Jet Propulsion Laboratories (JPL), Pasadena, CA designed and built named Explorer 1 (lasted 10 years!). Six months later, July 29, 1958, NASA, a civilian Agency was established By President Eisenhower.

In May 26, 1961, President John F. Kennedy declared a new Space goal for the USA. "By the end of this decade mankind will land on and return to Earth from the moon, not because it was easy, but because it is hard"! That kicked off the USA Manned Space Program, first with Project Mercury. Then Gemini and of course Apollo! Mercury 7 astronauts were selected from the trained military experienced flyers; US Navy, US Air Force, US Marines!

From one-man capsules to two man to three-man Apollo (conceived by the Eisenhower Administration) and many failures that took the lives of astronauts on the ground (January 27, 1967, Apollo I, Gus Grissom, Ed White and Roger Chaffee). Success was still a faraway goal.

Pure Oxygen atmosphere did not help them, neither did the highly flammable wires and materials inside the capsule.

During this timeframe, I was working at Sanders Microwave in New Hampshire and we received our first NASA Request for Proposal (RFP) out of NASA Goddard Space Flight Center (GSFC) in Greenbelt, MD.

I was given a copy to prepare a quote. Impossible, was my reaction after reading it. No company on the Planet would take this job! The requirements were out of this world…too rigid! So, we no bid that job!

It was twenty years later, until I found out the inside story on that first NASA bid out of GSFC!

At the time, GSFC was a spinoff of many US NAVY personnel so, after 90 companies' refusal to bid new jobs, they went back to Mil-Specs as their future requirements. That requirement the industry was familiar with and the contractor community accepted the new realistic specifications.

As one can see by all this history, the road to success did not happen by accident! Or perhaps it did as we shall soon see!

Next World news headline read, John Glenn, US Marine Corps Aviator on February 20, 1962, circled the globe three times, in 4 hours and 56 minutes at speeds of 17,000 miles per hour on Mercury capsule Friendship 7. See it displayed at the National Air and Space Museum in DC.

Exhibit MS – 9: President Eisenhower and Dr. von Braun with Col. Eifler (later LTG -3 Star General), et al, enjoying NASA Marshall Space Flight Center (MSFC) dedication on July 1, 1960. As a US Army Nike Ajax/Hercules Missile System Internal Guidance Repairman, assigned to the Ordnance Guided Missile School, I marched in that parade on a great weather day in Huntsville, Al.

I strongly recommend that you visit, as I did on my 50th anniversary tour of the Southland, the new **Space City Museum** in Huntsville for an unforgettable visit to the home of the USA Space Program.

Next for a foundation to this work, let's look at Space failures during this 40-year period of Space exploration. What you will see is how all the new measures we covered paid off big time!

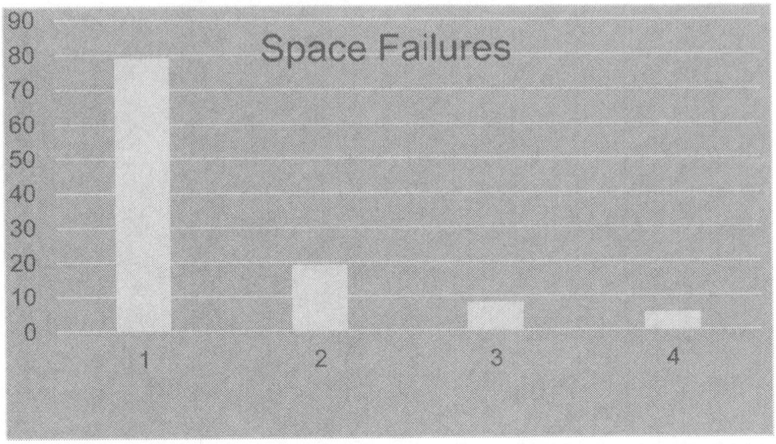

1957-1967 1968-1978 1979-1989 1990-2000

Exhibit MS-10: USA Space failures by Decades*

<u>" Space Systems Failures"*</u>, Harland & Lorenz, 2005

As one can see from Exhibit MS – 10, Space Failures fell way down as Launch Operations and payload designs that took in consideration space environment, launch speed, shock and vibration conditions were pretested by part suppliers, sub-systems and end item full-up end to end system testing.

Let's look at those leaps forward in Space Reliability over the almost 50-year timeframe.

The biggest payback for the US taxpayer out of DC Fed Gov, in my opinion, occurred when Mr. David Packard…. yes, the Packard of Hewlett-Packard Company was appointed as US Deputy Under Secretary of Defense in 1968.

What was called the **Packard Years** (1968-1971) in DC, Mr. Packard was familiar with electronic design and manufacturing. Being one of the first Silicon Valley ICON organizations, he analyzed the poor Quality and Reliability of military products and acted to, once and for all, fix the problem. Packard instituted what at the time was a "fly before buy" requirement. This had profound effect on the entire USA high-technology industry.

What it meant was that the producer of the product had to prove, in his plant, that the product met all specifications prior to shipment and after that payment occurred. This was during the Viet Nam Conflict and the guns and butter economy was flourishing.

At that time, I was employed by a top Naval electronic countermeasures (ECM) supplier. Guess what? They were faced with a "cannot ship" situation, due to failure of Systems Reliability requirements.

I was asked by my Savvy boss, who knew my hands-on background, if I would like to take a Reliability Management position and lead the effort to fix our out of specifications hardware. There I was without a degree, at 27 years old holding the fate of a company in my hands, in fact, many US Navy carrier pilots also, now that I think of it

Part-time, on the side, I was also enrolled full-time as a student at Northeastern University in pursuing my undergraduate BS degree on the new recently approved and, signed in 1961 by LBJ, the Cold War GI Bill ($2.5B for 5 years est.). And, I had nearly 10 hands-on years in the Electronics business (from micro-electronics to microwaves technologies) several at good old HP!

As I mentioned most failures at the time occurred with old-style first-generation technologies that failed when exposed to temperature and shock/vibration testing. Same happened here, I was familiar with HP Associates, a new solid-state house that Bill and Dave started on Page Mill Road in Palo Alto, CA, one of my favorite places. They were pushing the state-of the-art with micro-electronics devices and after a few phone calls got us started on fixing our systems. In addition, I had the pleasure of having two new team members who just graduated from Rensselaer Polytechnical Institute with their PHDs in EE join my team.

With the new HP solid state devices alone, we achieved success. Demonstrated systems reliability went from single digits to over 100 hours at worst case environments. It was rumored that those Carrier pilots refused to fly their missions unless they had our new technology on board and how you could blame them those SAM Ds were deadly once launched! Later over Kosovo, it was said that the USAF pilot flying a NATO plane did not have our ECM and was shot down. Years later, a Hollywood movie was made on this incident.

In addition, after creating computer circuit models of our systems, at my request, our new PHD EEs, found many unprotected first-generation solid-state circuitries susceptible to transients (once doing transient circuit analysis) thereby further increasing our ECM reliability. I first saw the problem at HP with those new Stanford EE designed circuits failing pre-maturely. Protective circuits were now added to existing designs as a preventative measure.

One of my team stayed on and became corporate Reliability Director, perhaps he is still there doing his thing to improve product offerings. I certainly hope so! I was off to other places and challenges once getting my AS, BS and MBA ending up at NASA for almost 20 years for my last hurrah!

Space Transportation System (STS) – the Space Shuttle - 1977- 2011

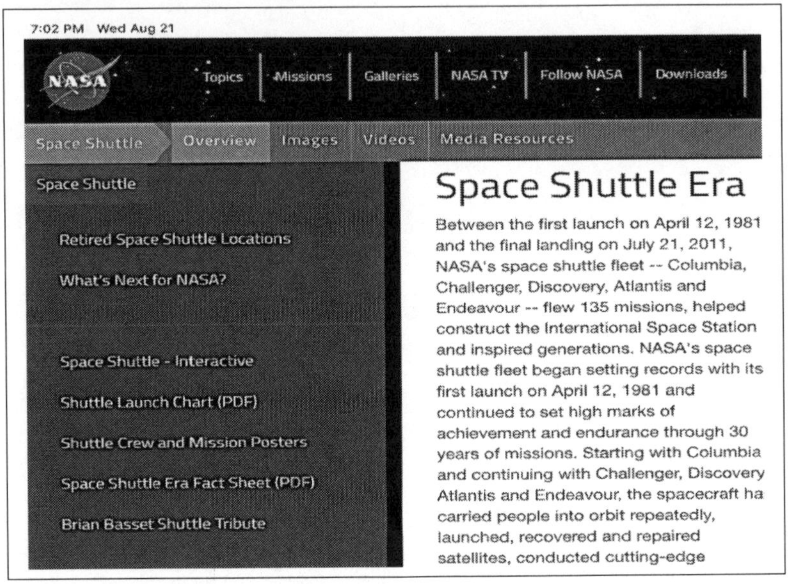

Exhibit MS-11: Space Transportation (Shuttle) Era.

Forth USA Space Missions: The Space Shuttle (STS) – 1977-2011

The first test flight occurred February 18, 1977, Columbia (STS-1) launched 5 years later April 12, 1981. STS-135 Atlantis retired the fleet on July 21, 2011. Six shuttles were built, the Enterprise was for flight test only. The others, Challenger, Endeavor, Atlantis, Discovery and Columbia all did many successful missions.

Description: Re-usable Space plane with External Tank and Solid Rocket boosters – 135 Flights!

Purpose: To act as space vehicle to transport large scale payloads to space on long duration missions

Results: over 135 launches conducted delivering many payloads for space exploration and science and communications purposes. A total of 355 people flew on Shuttles representing 16 countries.

Two catastrophic losses: Challenger & Columbia, and entire USA Astronaut Corps crews, 50% of the fleet! Both due to process flaws, first in design and application, second to violation of specifications. Results: 14 Astronauts killed, an unthinkable space scenario!

Lessons Learned: Space missions are still very risky and expensive, however, the paybacks to humankind are necessary trade-offs in achieving our Space mission goals and objectives.

Program achieved many amazing scientific and economic goals resulting in Mission Success.

Exhibit MS-12: Atlantis Shuttle, a workhorse of the STS Fleet, going out to pasture at KSC!

The STS system provided the USA and other fortunate countries, with a spacecraft low-orbit capability for 30 years and 135 flights. Besides deploying NASA spacecraft payloads, it was also used for 20 commercial satellites such a Telsat, SBS, ACS 7 & SCT. After the Columbia disaster, the plan to continue maintenance after the year 2010 was discontinued.

Where are they now? You might be asking….well, here is the answer:

https://www.nasa.gov/topics/shuttle_station/features/shuttle_map.html

1. **Endeavor – California Science Center, LA, CA**
2. **Atlantis – KSC Visitors Center Florida**
3. **Enterprise – Intrepid Carrier NYC, NY**
4. **Discovery – Smithsonian/Udvar-Hazy Center @ Dulles Airport near Washington DC**

STS returns to KSC at nighttime. One of many landings that occurred sundown with reliable results.

MS-13 STS Night Landing at KSC

Let's look at the Shuttles magnificent missions, some of which I watched in awe, every time we, at GSFC, had a payload on board. One was the HST payload launch from

KSC launch pad (LC-39B) on April 24, 1990 on Discovery STS-31. The average US citizen needed to be there to understand the difficulty of placing a Space Telescope the size of a Greyhound bus Strata Cruiser into space.

Hidden inside the shuttle payload bay, HST went for ride that took it to its new home 300 miles into space. Outside the Mission Control Center, we watched the shuttle climb swiftly and successfully placing HST into its new home in space. A magnificent site to behold and major accomplishment in USA Space history. HST is still performing after almost 30 years in service and five successful Servicing Missions. My GSFC Team covered the first four successfully and we received over 36 NASA medallions for our work contributions.

Exhibit MS – 14. Hubble Space Telescope (HST) Space Shuttle Crew responsible for on-orbit repair of telescope…autographed "Chuck O'Boyle: Thanks for your support for our successful flight"! STS Crew! My Team, including me, were honored with 36 NASA Medals for this successful mission work!

Exhibit MS – 15. Hubble Space Telescope First Servicing Mission NASA Medallion awarded to me for On-Board DF224 Computer Study conducted on fast track 30-day deadline prior to launch of STS-61 Endeavor Shuttle on December 2, 1993 to December 31, 1993.

Exhibit MS – 15A HST poster signed by STS-61 crew members, et al. at GSFC Celebration Party.

Chapter 5. MS Personnel Training/Certification

Ed. Note: Why add this subject to the Space Program, you may be asking? Because when you come right down to it people don't like losing. It is management's job to not only make a profit, and achieve goals/objectives, but to do it consistently, not only in the short run, but in the long run also. This requires building the right culture. Space training also, needs to focus on how this is done in difficult, high-risk situations. Nothing is worse than an un-informed manager, thus the reason for providing the reader here with some new knowledge and awareness…called Mission Success!

For almost 20 years at NASA, my job and my team's job focused on one goal…. achieving Mission Success! And at NASA, Management provided the culture that inspires employees to envision the impossible, perform with excellence and build incredible products.

To say the least, this was not an easy goal, however, in Satellite mission after mission, close to 50 by my count, we attempted to achieve that goal over and over. Why is the big question…how, when, where and why answers follow?

Hea Chuck! What about NASA Challenger, Hubble, Columbia and other failures? Yes, we will discuss them here. But all were not with Goddard Space Flight Center (GSFC) missions! I arrived after the Challenger Accident (KSC mission), Hubble was done at Marshall Space Flight Center (MSFC) in Huntsville, AL. GSFC was handed it off after launch and on-orbit operation. I supplied my team to support four HST Repair Missions that were all successful. And Columbia was another KSC mission.

In the Presidential Commissions Report on the Challenger Disaster, I went deeper into hearings reports on file at the GSFC Library (Great one!). One hearing had Phil Crosby, founder of the Quality College in Winter Park, Fl. Phil was VP Quality at ITT for many years working for Harold Geneen, CEO ITT and Business guru of the 20th Century. He was not too happy that NASA at the time laid off almost 1500 Quality personnel. Severely undermanned, checks and balances were out of control according to Crosby. One of the reasons I was hired by NASA GSFC was to fix this serious problem, in short order or else, and I did at GSFC and KSC, where I had personnel. I had personnel at 15 locations across the USA and on-site at GSFC & KSC. Our bases were covered by experienced, hands on aerospace professionals. And all this without computers. For at the time I was told if I or any of my personnel were seen using a computer, we would be all fired. Can you imagine that? In today's NASA workplace if you did not know how to use a PC you would be fired! "Computers were for secretaries, he screamed!" He demanded that all personnel have eyes and ears on the hardware. We were to be a second look, he continued! But what about Dave who does your admin work? Oh, he is an exception, he replied!

Such was the scene at the time. You had entrenched power players who feared the computer for years maintaining their kingdoms from foreign objects. Upon his retirement years later, we bought over 100 desktops and were forced to use Microsoft Office instead of the Apple and superior software we were all used to at our other jobs. Today, I doubt if you will not find everyone using PCs at their workplace, for you cannot do your job without one. I too, took all the training courses on Microsoft Office suite and Project Manager too!

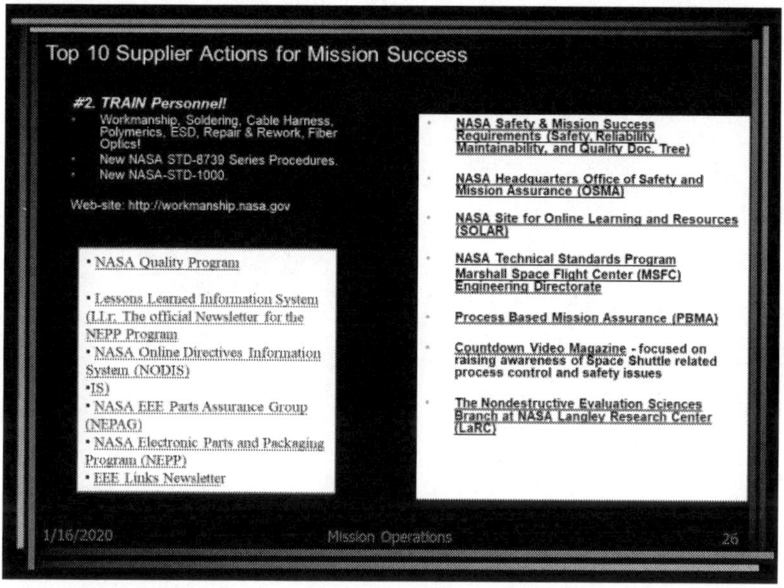

Exhibit MS – 16: Listing of NASA Training Requirements.

Several GSFC satellite missions did have problems and we will cover them here. I participated in the Failure Review Board on LandSat 7 and our Report is online for further review.

Well, it takes many resources to achieve Continuous Mission Success at NASA GSFC and all aerospace contractors.

First and foremost, it was the team of players (personnel resources), all trained by NASA Training Programs (See Exhibit MS – 16) in such skills Workmanship Standards (Soldering, ESD, Repair & Rework, etc.) …. knowing the right way, the NASA way, of doing the job. This embedded a common purpose of doing the job right the first time. Annual re-certification was also a requirement to maintain skills. And followed by more training in the Quality Control function.

We even added a two-part course in Flight Assurance Management I & II that I helped develop and even thought to our entire Directive.

NASA GSFC Flight Assurance Manager's (FAM) Training Course Phases/Volumes I & II.

This was my second assignment upon arrival at Goddard. The first phase, **Volume I**, covered the basics of Quality Control and Assurance. From the Western Electric studies to the Japanese post WWII Deming taught lessons on Quality; we gave an in-depth understanding of the principles and practices that were necessary to accomplish what NASA called Mission Success.

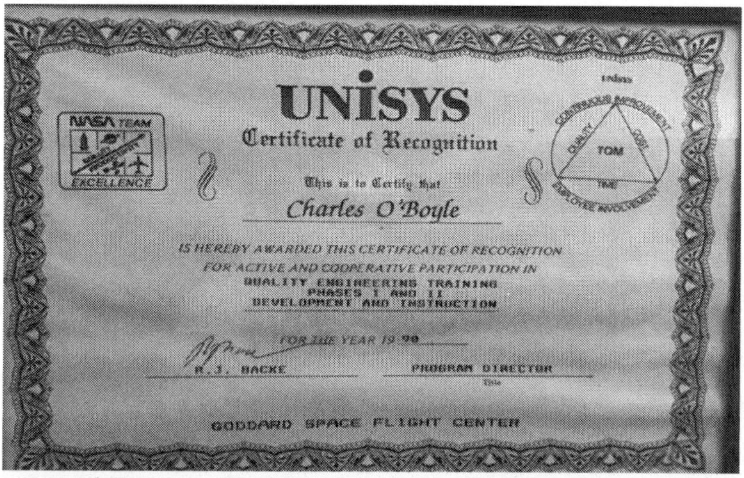

UNISYS

Certificate of Recognition

This is to Certify that

Charles O'Boyle

IS HEREBY AWARDED THIS CERTIFICATE OF RECOGNITION
FOR ACTIVE AND COOPERATIVE PARTICIPATION IN
QUALITY ENGINEERING TRAINING
PHASES I AND II
DEVELOPMENT AND INSTRUCTION

FOR THE YEAR 19 90

R.J. BACKE — PROGRAM DIRECTOR

GODDARD SPACE FLIGHT CENTER

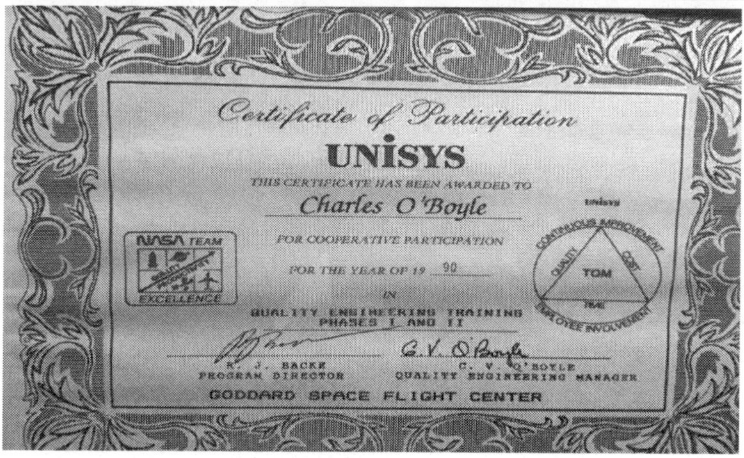

Certificate of Participation

UNISYS

THIS CERTIFICATE HAS BEEN AWARDED TO

Charles O'Boyle

FOR COOPERATIVE PARTICIPATION

FOR THE YEAR OF 19 90

IN

QUALITY ENGINEERING TRAINING
PHASES I AND II

R. J. BACKE, PROGRAM DIRECTOR — C. V. O'BOYLE, QUALITY ENGINEERING MANAGER

GODDARD SPACE FLIGHT CENTER

Exhibits MS - 16 A&B: Certificates of Recognition & Participation Part I & II Phases

Second phase, **Volume II**, used all the Flight Assurance Manager (FAM) Task Leader's knowledge of how their individual programs implemented the many NASA Standards and requirements that were embedded into every mission's program plan. In addition, the spacecraft technology was covered in depth. And each mission's on-orbit goals and objectives were communicated by the NASA Goddard Space Flight Center Project's individual Team Leaders. Two large volumes of training, study and reference material and instructor overhead slides were prepared for this highly successful training program. It was repeated for several years after, however, as with many classics and with new management, this too, was tossed by the wayside for the new flavor of the month training fad.

We were viewed by management as a "second look", eyes and ears for NASA wherever our work assignments took us from all the way back to the suppliers' plants and right up to the launch pad and even after launch at mission operations. After Columbia, even a third set of eyes was involved, experts in Risk Management (RM) joined the mission teams with their assessments and recommendations.

Secondly, it was the system and its documentation from Standards to Handbooks (like the Golden Rules) to Military specifications on parts and processes, all worked as barriers (like jersey walls on the highway) to ward off failures.

Lastly, but not least was a third, especially for Space Flight all systems were designed by engineering for triple redundancy. Just like your car has an emergency brake should that brake pedal not stop you…you hit the emergency brake to safely bring you to a stop. In the old days you could always downshift to slow you down, this is like triple redundancy, built right in your car.

All along the design, fabrication and build process we had a trust but verify process that permitted advance to the next station. Full-up, testing, end to end as we called it as closely as possible to the real space environment was required and verified on process travelers throughout the mission up to before and after orbit certification. Test, test, test was the mantra voiced by all acceptance personnel.

Requirements and documentation are part of the mission success process. Such references as Systems Engineering Management Plans and Risk Management Handbook are embedded into Program/Project Plans. These are constantly revised as lessons learned are applied to NASA Standards. For example, in April 2010, NASA issued NASA/SP-2010-576, the NASA Risk-Informed Decision-Making Handbook.

In addition, each new Project is required to have a tailor-made Office of Flight Assurance PAR now called a MAR (Mission Assurance Requirements). This is a high-level Specification that details what I call the toll gates for compliance for flight acceptance. See Exhibit MS – 48, for MAR contents, deliverables and mitigation factors.

Program Design Reviews were a continuous part of the verification process: technical meetings such as: SRR, PDR, CDR, FRBs, SMRTs, etc., are all programmed in the master schedule of formal engineering tasks.

As we matured with our Lessons Learned Data Base, we placed mission **success factors** right into the requirements and specifications of any new satellite builds. A recent conference paper provides a good look at first generation vs present generation successes at NASA. I recommend reading it to see the differences involved.

"Success Factors in Human Space Programs - Why Did Apollo Succeed Better Than Later Programs?" By: Harry

W. Jones, 1 NASA Ames Research Center, Moffett Field, CA, 94035-0001 at the: 45th International Conference on Environmental Systems ICES-2015-048, 12-16 July 2015, Bellevue, Washington

Chapter 6. MS Program/Project Management Planning

Let's look at some training techniques used in US Army training schools to begin this chapter.

I first was exposed to them at Fort Monmouth Basics Electronics School. We call them memory joggers or ways to remember long concepts.

The first one is the letters DMAIC for Define, Measure, Analyze, Improve and Control. So, lets apply the technique for planning.

Define – the projects scope, objectives, resources and constraints.

Measure – groups, CCCR (reviews), obtain data for Quarterly meetings.

Analyze – analyze data to determine root causes.

Improve – Intervene to improve performance. Plan new goals.

Control – implement a control system to maintain performance over time.

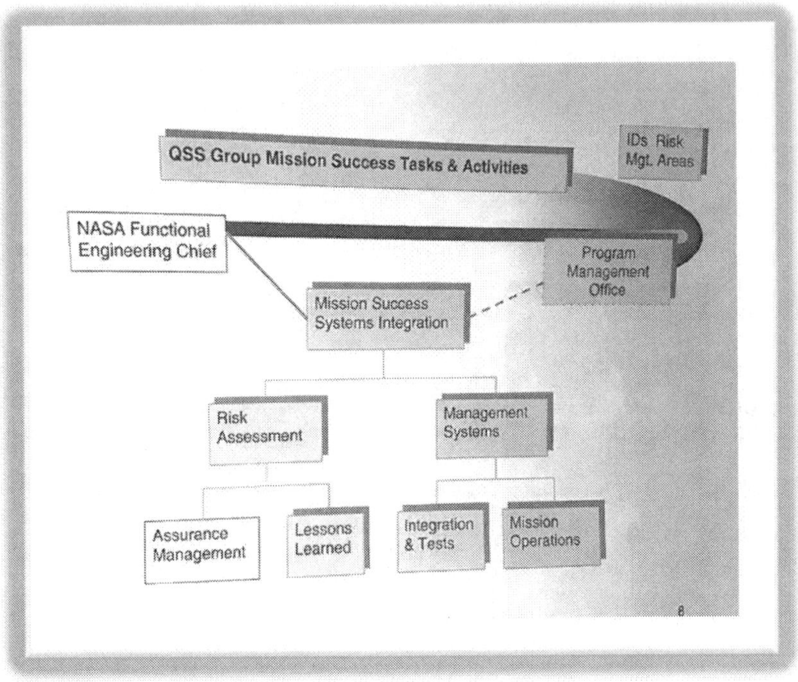

Exhibit MS 17: Mission Success Tasks & Activities

Exhibit MS – 17, shows how the Mission Success Tasks and Activities are integrated and interacts with existing requirements. Here again is the matrix management organization structure and how it relates to the Program Management Office (PMO). All activities are part of a one team focus that once embedded with MS practices, achieve a common mindset of achieving success.

The PMO has the responsibility for laying out detail plans for all Spacecraft design, build and fabrication efforts. The following NASA Website covers the NASA program management policy, operations and requirements in detail. So, you too can visit it for more details.

(https://www.nasa.gov/offices/oce/functions/prog_proj_mgmt.html).

As we discussed earlier, we saw Mission Success as an embedded Engineering entity like Quality, Reliability and Safety disciplines, whose focus was on planning, analysis and navigating continuous improvement in all Project activities.

"At NASA, the NASA Office of the Chief Engineer (OCE) establishes Agency-level program and project management policies and requirements and provides guidance to assist the programs and projects in the implementation and tailoring of the policy requirements."

"It is NASA policy to apply program and project management procedural requirements, to plan, track progress, accomplish mission objectives, and follow the requirements and life-cycle reviews unique to their investment area as defined in the NASA procedural requirements for space flight, research and technology, information technology (IT) and institutional infrastructure".

"The procedural requirements establish the framework and process by which NASA formulates and implements programs and projects consistent with the governance model NPD 1000.0, NASA Governance and Strategic Management Handbook, and establish a standard of uniformity in the management of programs and projects".

The OCE is responsible for the overarching Agency program/project management policy:

- NPD 7120.4 — Program/Project Management: The following organizations are responsible for the program and project management procedural requirements related to each of the four investment areas indicated below:

- Office of the Chief Engineer (OCE): Space Flight, and Research and Technology:
    - NPR 7120.5 — NASA Space Flight Program and Project Management Requirements
    - NPR 7120.8 — NASA Research and Technology Program and Project Management Requirements
- Office of the Chief Information Officer (OCIO): Information Technology:
    - NPR 7120.7 - NASA Information Technology Program and Project Management Requirements
- Office of Strategic Infrastructure (OSI): Institutional Infrastructure (includes Facilities Projects and Environmental Programs):
    - NPR 8820.2 - Facility Project Requirements (FPR)
    - NPR 8590.1 - Environmental Compliance and Restoration Program

Additional information on Program and Project Management (PM) may be found on the PM NEN Community of Practice site at:

https://nen.nasa.gov/web/pm

and the OCE Tab in NODIS at:

http://nodis3.gsfc.nasa.gov/OCE_rep/OCE_list.cfm.

For reader reference purposes, I have lifted just the planning applicable NASA Requirements (NPR 7120-10A) Documents and Forms Chapter contents from the NASA.gov public Website. As you can see, they are quite detailed and provide the basis for managers to use on every spacecraft project.

COMPLIANCE IS MANDATORY: Note: NASA will verify compliance to these requirements.

NASA Applicable Documents and Forms:

a. NPR 1441.1, NASA Records Management Program Requirements.

b. NPR 7120.5, NASA Space Flight Program and Project Management Requirements.

c. NPR 7120.7, NASA Information Technology and Institutional Infrastructure Program and Project Management Requirements.

d. NPR 7120.8, NASA Research and Technology Program and Project Management Requirements.

e. NPR 8715.3, NASA General Safety Program Requirements.

f. NPR 8900.1, NASA Health and Medical Requirements for Human Space Exploration.

g. NASA-STD 8709.20, Management of Safety and Mission Assurance Technical Authority (SMA TA) Requirements.

Another NASA web-site worth noting for this MS Chapter is the following:

NASA APPEL: appel.nasa.gov
Through its world-class training curriculum, knowledge-sharing initiatives, development resources, and strategic communications, APPEL Knowledge Services helps ensure NASA's technical workforce has the skills and knowledge needed to advance mission success.

Additional NASA Resources available for all Project Managers:

<u>NASA's Project Management Handbook</u> (NASA/SP-2014-3705 – Sept. 2014) is available on-line at the following NASA Website:

(https://ntrs.nasa.gov/archive/nasa/casi.ntrs.nasa.gov/20150000400.pdf)

This NASA PM Handbook covers managing the complete life cycle of a mission and all the details of developing a Work Breakdown Structure (WBS) that is a must to control all aspects of a new project.

In addition, a NASA Schedule Management Handbook helps the manager by providing additional details of schedule control and schedule reporting.

Chapter 7. MS Program/Project Staffing

In order to get a Flight Assurance job at NASA you had to beat the odds of getting into the Naval Academy at Annapolis, MD. My HR guy gave me the statistics one day…one in 1500 resumes is what he told me. And I believed him since I sent in my resume twice before an interview was requested. Mine came on a Sunday afternoon and was conducted by a 35-year NASA veteran almost going back to day one. I had almost 30 years in the business at the time and went up the ladder from Technician to Engineer to Manager to Director. My interview came right after the Challenger disaster and NASA investigation committees scolded the Agency for letting go so many Flight Assurance personnel. After just a few questions, he asked when can you start? Later, I found out that I was carefully pre-checked before the interview verifying my "track record"! I had Fed Gov Clearance since I was 18 years old, so there was no problem there.

My job was simply stated by my NASA boss. I want you to find me the best people on the planet. I don't care what you pay them or where you get them…but get them fast ok? Essentially what he was looking for was the next generation of NASA workers, highly experienced and over-qualified if necessary. Fortunately, for both of us the USA high technology was in a slow down pace. The West Coast seemed to be suffering more than other places with the airplane industry consolidation. Within the next several months I was able to hire dozens of highly qualified new hires from all over the USA. Today many are still at NASA and many went on to higher level positions even to the Director level. I always hired people I thought were smarter than myself! Several hires were retired managers from the best USA companies. They still had "fire in the belly" and what it takes to perform the critical tasks they were assigned to perform.

At NASA, as in industry, teamwork was a very important success factor. We did not look for prima donnas, we looked for people who got results and were dedicated to the cause of mission success. All hires were seasoned professionals in the field of Quality Assurance which is what the new openings required.

Now my Team of Flight Assurance experts was up to full billets. Assignment was then fitted to a mission functionality. For example, those team members with microwave were assigned to Satellite Programs like TRMM (Tropical Rainfall Measurement Missions) that had 17 years of performance in space! Others like HST SM1 required an optical specialist. We covered that with a Princeton alumnus with 50 years' experience! They also needed a space tools specialist for performing space walk repairs. I found an ex-Canadian Aerospace Quality specialist who did a bang-up job that received many commendations for his effort.

At KSC I transferred our first female inspector who became a valuable Launch Team player for many years. Another two female inspectors who completed satellite training with glowing marks were added to bread and butter satellite missions such as Explorer and Hitchhiker.

16 NASA Contractor locations throughout the USA were also covered by my Team. I performed round robin trips monthly to provide coaching to all NASA delegates there.

They also came back to their home base at GSFC for training and re-certification to NASA Workmanship Standards.

As one can see mission success did not happen by chance or accident, careful planning was always involved to fit the right skills and experience with the mission's needs and goals.

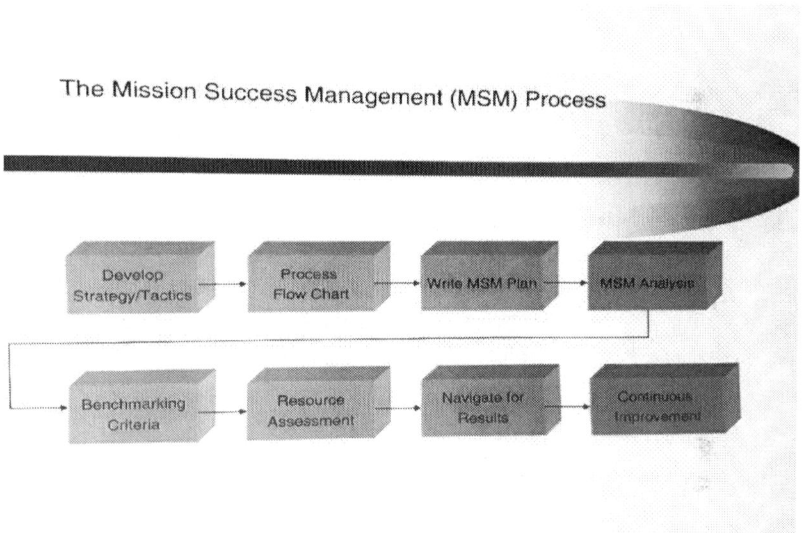

Exhibit MS – 18: Mission Success Management (MSM) Process

MSM consist of an <u>8-Step</u> activity:

1. Develop a tailor-made project unique Strategy & Tactics to be used in the MSM Plan.
2. Lay out the processes from Design to Fabrication to Build to Inspection & Test to final delivery and launch/post launch certification.
3. Write the MSM Plan incorporating Step 1 & 2.

4. Conduct an MSM Plan Analysis on Quality, Reliability, Safety & Risk.
5. Start using Benchmarking as a standard of comparison and lessons learned.
6. Perform a Project Resource Assessment using success factors.
7. Navigate for Results: use milestones, schedules, costs results.
8. Perform Continuous Improvements over all activities.

It is the responsibility of the Project Manager (PM) and the Flight Assurance Manager (FAM) to place this 8-step activity into the Master Schedule. Assurance Management then assists in continuing to monitor compliance to requirements throughout the life of the Program.

Chapter 8. MS Management & Organizing

I am very surprised when I look at our governmental and private businesses how little thought is given to correctly organizing a large group of technical personnel.

For example, I recently reviewed a so-called world class auto producer and found that 48 VPs were involved at their corporate headquarters. Hell, we fought WWII with less Generals…I recall only nine 5 stars: 4 Army, 4 Navy and one Air Force.

Chrysler, almost bankrupt when CEO Lee Iacocca took over, had 33 VPs on day one. Day two only one survived. Lee fired 32 VPs in one day and turned around a disastrous situation in just two years.

Way back a fellow named Henry Fayol (1841-1925), Father of Scientific Management, a mining executive, conducted deep studies (book entitles ***General & Industrial Administration*** 1925) as to the 5 functions of management and 14 principles applicable to technical and managerial activities. Today these principles are taken for granted when viewed by the novice in organizations. Plan, Organize, Direct, Staff and Control is what managers are supposed to do. That defines their job. And the 14 principles are how to do it! I will not try to top that, but we can use them to design our optimum organization.

1. Division of work
2. Authority
3. Discipline
4. Unity of command
5. Unity of Direction
6. Subordination of Interests

7. Remuneration
8. Centralization & Decentralization
9. Scalar Chain
10. Order
11. Equity
12. Stability of Tenure
13. Initiative
14. Espirit de Corp

Let me asks you pros now, honestly, how many of you have never seen this before…my guess is not many! I discovered it by accident trying to deal with almost 100 direct reports and their annual reviews as an Engineering Manager at NASA as a contractor. Hell, even the head bean counter, one hell of a nice guy, teased me, Chuck they will have hundreds of personnel reporting to you if you don't complain. So, I did some research and discovered Fayol's book! Alas, I immediately wrote a memo to my boss and provided a new organization chart of my direct reports. The big change was appointing Team Leaders by project and this one step solved my organization problem. Oh yes, no one had more than the 12 personnel maximum rule that Fayol proved out. I think someone else, centuries ago, had 12 direct reports also and it started a worldwide movement still existing thousands of years later!

Now with that foundation, any manager worth his or her salt, can set up an effective organization. Most will not need to as most on-going projects have been around for years evolving with new tools and technologies. At NASA, almost all programs and projects had the classic Program Management structure. This is made up of essential functional members from Engineering/Design/Systems,

Fabrication, Administration, Quality, Reliability, Safety & Contracts.

From an academic point of view, NASA is a classic example of matrix management in form and structure. At the top exists a corporate like structure with headquarters offices offering Space expertise to the line activities that are space function specialties and are called Space Centers. Along with Goddard (GSFC) there are 12 other space specialties located across the USA. A top level view of Headquarters organization is seen at the following web-site, (https://www.nasa.gov/about/org_index.html). The Center locations and their specialty areas are shown in the following web-site (https://www.nasa.gov/about/sites/index.html).

Using this matrix management style of organization space specialists are assigned to various programs or projects as their expertise are needed. This organization flexibility offers functional specialist from Engineering, Quality (SRMQA), Fabrication, Logistics, Operations when needed, at the right time and place the various project needs them.

NASA's new handbook "***Program and Project Management Handbook***" provides complete guidance for the mission manager to utilize in running a Spacecraft mission.

(https://www.nasa.gov/pdf/423715main_NPR_7120-5_HB_FINAL-02-25-10.pdf)

Chapter 9. MS Project Controlling

In his classic Management book, Peter Drucker cites that "the less complex a business is, the fewer things can go wrong. And the more complex a business, the more difficult it is figure what went wrong and to take the right remedial action." We engineers are so familiar with Murphy's Law, if anything can go wrong, it will. Drucker has his own Law, "If one thing goes wrong, everything else will, and all at the same time"! The prevention is knowing one's own business to deal with the wrongs is the answer. Nothing proves this better than the Apollo 13 mission, and Gene Krantz' statement that, ***Failure is Not an Option***! He wrote a book with this title; I have an autographed copy!

Exhibit MS – 19: Three NASA Amigos with Gene Krantz, Apollo 13 fame, movie & book, *"Failure is Not an Option"* mantra. L to R: NASA, since the Apollo mission, Joe Osche, Gene, Jim Suraci & Me.

PS Shortly after this photo Gene (a fourth degree K of C member) contributed all his NASA Memorabilia to his local church and they were lucky he did! For years later, he lost all his possessions and home in the disastrous hurricane flood in Houston, TX. As I write this book, Gene is alive and well still doing the celebrity circuit and wishes me well on my new book.

As shown in the block-buster Hollywood movie, that only "Right Stuff" actor, Ed Harris, could play as good a part as real-life NASA Mission Control Director Gene Krantz. That event emboldened the can-do spirit of NASA at the time. Of course, we are also talking about Korean War veterans such as Gene Krantz who made those times unforgettable to us all in the USA.

Now for the real story: according to Wikipedia…

“As Gene described his statement to Hollywood movie producers, In preparation for the movie, the script writers, Al Reinart and Bill Broyles, came down to Clear Lake to interview me on "What are the people in Mission Control really like?" One of their questions was "Weren't there times when everybody, or at least a few people, just panicked?" My answer was "No, when bad things happened, we just calmly laid out all the options, and failure was not one of them." ... I immediately sensed that Bill Broyles wanted to leave and assumed that he was bored with the interview. Only months later did I learn that when they got in their car to leave, he started screaming, "That's it! That's the tag line for the whole movie, Failure is not an option!”

The History Channel has a real-life documentary on the incident. Recreating that time in Space history was challenging to say the least. But both the documentary and movie are must see viewing for all Americans, in fact all the planet for that matter!

AS shown in Exhibit MS – 20 below, NASA’s Mission Control Center (MCC) is where it is at as far as monitoring space events. On shuttle launches it begins in Florida at the Merritt Island Tracking Center (MILA) until the launch clears the tower. Then, transfer goes to Houston seen below until the mission is over.

Exhibit MS – 20 NASA Mission Control Center

By NASA/Bill Ingalls - http://spaceflight.nasa.gov/gallery/images/station/crew-9/html/jsc2004e19918.html, Public Domain, https://commons.wikimedia.org/w/index.php?curid=241586

A NASA **Mission Control Center** (**MCC**, sometimes called a **Flight Control Center** or **Operations Center**) is a facility that manages space flights, usually from the point of launch until landing or the end of the mission. It is part of the ground segment of spacecraft operations. A staff of flight controllers and other support personnel monitor all aspects of the mission using telemetry, and send commands to the vehicle using ground stations. Personnel supporting the mission from an MCC can include representatives of the attitude control system, power, propulsion, thermal, attitude dynamics, orbital operations and other subsystem disciplines. The training for these missions usually falls under the responsibility of the flight controllers, typically including extensive rehearsals in the MCC.

I had the opportunity of auditing all USA NASA Tracking Stations and Merritt Island was one of them. They all impressed me with their trained personnel some around from the start of the STS Program. This gave me added assurance that our Astronauts had the best support teams they could rely on for achieving mission success.

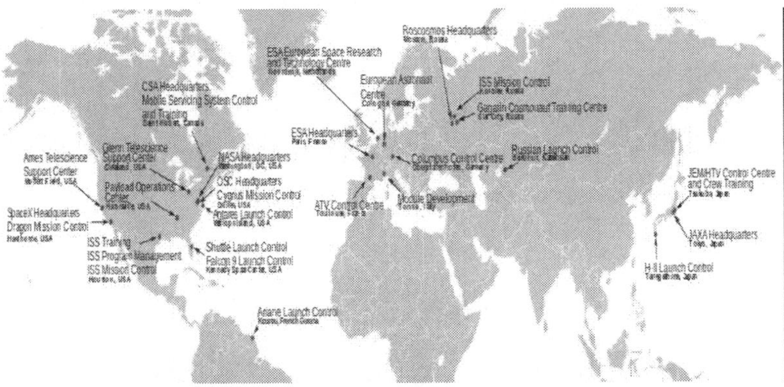

Exhibit MS – 21: International Space Station World-wide Tracking locations.

As shown in Exhibit MS – 21, for the Earth orbiting NASA International Space Station (ISS), a world-wide network of communications is necessary to maintain 24/7/365 coverage of on-orbit operations.

In addition to mission control operations, special missions have what is called Control Rooms that maintain continuous communications with the orbiting payload.

A control room's purpose is production control, and serves as a central space where a large physical facility or physically dispersed service can be monitored and controlled. Central control rooms came into general use in factories during the 1920s.

Control rooms for vital facilities are typically tightly secured and inaccessible to the general public. Multiple electronic displays and control panels are usually prese8nt, and there may also be a large wall-sized display area visible from all locations within the space. Some control rooms are themselves under continuous video surveillance and recording, for security and personnel accountability purposes. Many control rooms are manned on a "24/7/365" basis, and may have multiple people on duty at all times (such as implementation of a "two-man rule"), to ensure continuous vigilance. (https://en.wikipedia.org/wiki/Control_room).

Other special-purpose control room spaces may be temporarily set up for special projects (such as an oceanographic exploration mission), and closed or dismantled once the project is concluded.

As illustrated by Exhibit MS – 22 below, a NASA mission operations **ground segment** consists of all the ground-based elements of a spacecraft system used by operators and support personnel, as opposed to the **space segment** and **user segment**. The ground segment enables management of a spacecraft, and distribution of payload data and telemetry among interested parties on the ground. The primary elements of a ground segment are:

- 8Ground (or Earth) stations, which provide radio interfaces with spacecraft
- Mission control (or operations) centers, from which spacecraft are managed

- Ground networks, which connect the other ground elements to one another
- Remote terminals, used by support personnel
- Spacecraft integration and test facilities
- Launch facilities

These elements are present in nearly all space missions, whether commercial, military, or scientific. They may be located together or separated geographically, and they may be operated by different parties. Some elements may support multiple spacecraft simultaneously.

(https://en.wikipedia.org/wiki/Ground_segment)

At GSFC, all our missions are monitored by Missions Operations Center (MOC) located on site. It was here that made it possible for us to watch at home the moon landing 50 years ago. As communications were passed through vast networks to be translated to video by GSFC MOC Specialists.

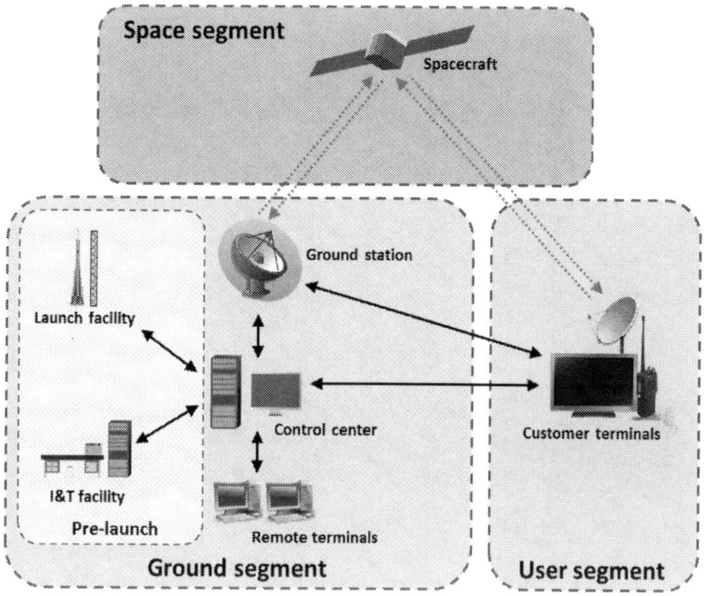

Exhibit MS – 22: NASA Ground Control Communications Interaction

Once NASA adopted AS9000, a similar checklist of requirements as shown below in Exhibit MS – 23, became the Quality Standard. In addition, AS9000 added 34 additional parts for the aviation industry to assure Safety, Reliability and Quality Standards were complied with.

ISO 9001 - Assessment Checklist (ANSI/ASQC Q9001-1994)

4.1	MANAGEMENT RESPONSIBILITY	4.10	INSPECTION AND TESTING
4.1.1	Quality Policy	4.10.1	General
4.1.2	Organization	4.10.2	Receiving Inspection and Testing
4.1.2.1	Responsibility and Authority	4.10.3	In-process Inspection and Testing
4.1.2.2	Resources	4.10.4	Final Inspection and Testing
4.1.2.3	Management Representative	4.10.5	Inspection and Test Records
4.1.3	Management Review		
4.2	QUALITY SYSTEM	4.11	CONTROL OF INSPECTION, MEASURIN TEST EQUIPMENT
4.2.1	General	4.11.1	General
4.2.2	Quality-system Procedures	4.11.2	Control Procedures
4.2.3	Quality Planning		
		4.12	INSPECTION AND TEST STATUS
4.3	CONTRACT REVIEW		
4.3.1	General		
4.3.2	Review	4.13	CONTROL OF NONCONFORMING PROD
4.3.3	Amendment to Contract	4.13.1	General
4.3.4	Records	4.13.2	Review and Disposition of Nonconforming Pr
		4.14	CORRECTIVE AND PREVENTIVE ACTION
4.4	DESIGN CONTROL	4.14.1	General
4.4.1	General	4.14.2	Corrective Action
4.4.2	Design and Development Planning	4.14.3	Preventive Action
4.4.3	Organizational and Technical Interfaces		
4.4.4	Design Input	4.15	HANDLING, STORAGE, PACKAGING, PRESERVATION, DELIVERY
4.4.5	Design Output		
4.4.6	Design Review	4.15.1	General
4.4.7	Design Verification	4.15.2	Handling
4.4.8	Design Validation	4.15.3	Storage
4.4.9	Design Changes	4.15.4	Packaging
		4.15.5	Preservation
4.5	DOCUMENT AND DATA CONTROL	4.15.6	Delivery
4.5.1	General		
4.5.2	Document and Data Approval and Issue	4.16	CONTROL OF QUALITY RECORDS
4.5.3	Document and Data Changes		
		4.17	INTERNAL QUALITY AUDITS
4.6	PURCHASING		
4.6.1	General	4.18	TRAINING
4.6.2	Evaluation of Subcontractors		
4.6.3	Purchasing Data	4.19	SERVICING
4.6.4	Verification of Purchased Product		
4.6.4.1	Supplier Verification at Subcontractor's Premises	4.20	STATISTICAL TECHNIQUES
4.6.4.2	Customer Verification of Subcontracted Product	4.20.1	Identification of Need
		4.20.2	Procedures
4.7	CONTROL OF CUSTOMER-SUPPLIED PRODUCT		
4.8	PRODUCT IDENTIFICATION AND TRACEABILITY		
4.9	PROCESS CONTROL		

Exhibit MS - 23: List of ISO 9000 Requirements for applying QMS Function.

Main items on the ISO checklist included:

1. Management Responsibility
2. Quality System
3. Contract Review
4. Design Control
5. Document & Data Control

6. Purchasing
7. Control of Customer Supplied Products
8. Product Identification and Traceability 18 St
9. Process Control
10. Inspection & Testing
11. Control of Non-conforming product
12. Corrective & Preventative Action
13. Control of Quality Records
14. Internal Quality Audits
15. Training
16. Servicing
17. Statistical Techniques

Exhibit MS – 24: Mission Control Center for Hubble Space Telescope (HST).

(Public Domain, https://commons.wikimedia.org/w/index.php?curid=4751874)

Exhibit MS – 24 shows the HST Control Center where 24/7/365 communications coverage takes place that deals with any anomalies that occur with the spacecraft or tracking stations.

One task my NASA Civil Service customer and I performed was a review of all NASA Tracking Stations in the USA. This involved visiting and evaluating on-site practices such as: documentation, maintenance, training, metrics and ISO compliance results.

I prepared a NASA Tracking Station Benchmarking Plan for our use in conducting these evaluations. And broke it down into the following sections:

1. Documentation Control
2. Process Control
3. Equipment
4. Testing
5. Personnel
6. Corrective Action
7. Assessments
8. Statistical Techniques
9. Safety
10. Quality Improvement

From this effort, we were able to provide a listing of Best Practices and Success Factors that was shared with all sites.

A listing below in Exhibit MS – 24A, helped provide key Success Elements for all Earth Science/Space Science

Mission Control Centers and Tracking Stations. Both Quality and Safety functions were covered.

NASA Tracking Station Operators, Facilities and Equipment were the major headings. Details are shown below:

Operator:	**Verification**	**Safety Checklist**
1. Training		**Personnel:**
2. Standards		Mgt. Commitment
3 Skills		Hazards Analysis
4. Job Knowledge		Prevention & Control
5. Certification		Safety & Health Training
6. Fatigue Factor		Workmanship Certification
7. Buddy System		Communications
8. Records		Emergency Training
Facilities:		Ergonomics
1. Use of Procedures		Reviews (LRR/FRR)
2. Use of Checklists		**Facilities:**
3. Maintain Equipment		Layout
4. Calibration		HVAC
5. Env. Guidelines		Electrical Backup
6. Backup Power		Mechanical Backup
7. Aging analysis		Pressure

8. Records		Process monitoring
Equipment:		Operations
1. Vintage		Documentation
2. Backup		**Equipment:**
3. Records		Maintenance Program
4. Calibration/ Maintenanc e		Certification Requirements
5. Upgrade Plans		Storage & Preparation
		Location & Use
		Records
		Backup Capabilities

Exhibit MS – 24A Checklist for Tracking Station Benchmarking Exercise.

Our visits were welcomed by all NASA Tracking Station personnel, it was one of the few times they said that they had interested people in their operations. I was impressed by their dedicated personnel existing at every site and the NASA Team spirit embedded in their operations.

Chapter 10. MS Project Directing

At NASA GSFC Program direction is the responsibility and authority of the Program or Project Manager. Team members are away from their functional organization management, but not divorced from their reach. This means that performance reviews are still conducted by their employer, if they are a contractor. Often his or her functional manager discusses the performance with each program management to determine if customer satisfaction is being met.

In any matrix organization structure constant monitoring is required so any problems can quickly be remedied. Using a paternalistic management style, team members are like family and many times the stars or top performers are sought out by new program managers long before a project begins.

The functional organizations have regular meetings to keep every member informed of latest news. Today, computerized cell phones and iPads are used taking the old beeper as a method of contacting a team member quickly. PCs are also used to communicate all aspects of a projects master schedule and acts as a roadmap for future activities. Today, there is no reason not to be informed of scheduled events as readiness meetings and major testing activities.

With the occurrence of Contractor accidents, a 24/7/365 video cam operation of spacecraft build activities is now a means of control of contractor progress. Daily telecons are now replaced by video reviews to keep the PMO informed of progress. With international partners, NASA Teams all over the planet now are up to speed on a spacecraft mission from design to launch!

And then there are the many contract milestones, some even for progress payments to the contract team. Common are PRD, CRD, LRR all major program reviews by management.

The work breakdown structure (WBS) is the roadmap that lays out the tasks and activities of the Project. All team members sign up to it and it is time-lined in nature. Each is covered from beginning to end…or design, build right up to launch and certification.

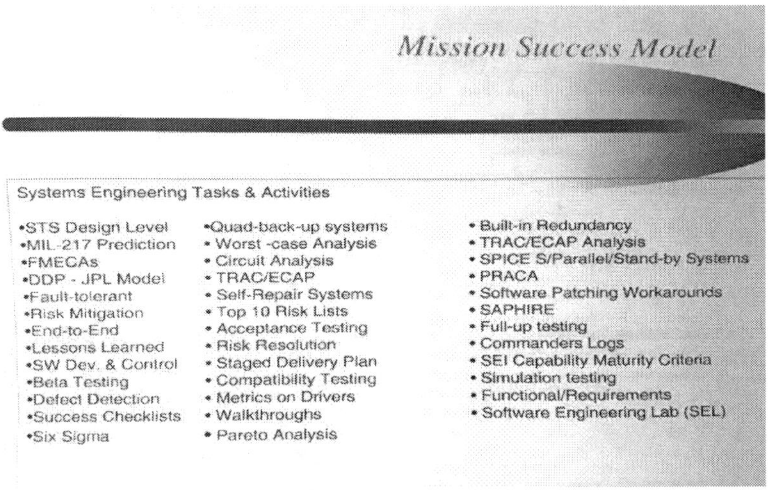

Exhibit MS – 25: Our Recommended MS Model: Tasks & Activities

As illustrated in Exhibit MS- 25, Our MSM covered all hardware and software tasks and activities starting with Reliability Predictions, Risk Mitigation, Trace/ECAP Analysis, Software Plans even Commander's Logs.

Thirty-Eight (38) engineering activities are listed in the above MS Model, all providing a strong foundation for spacecraft mission success. As new technologies become available, especially in the Software arena, additional requirements need to be added.

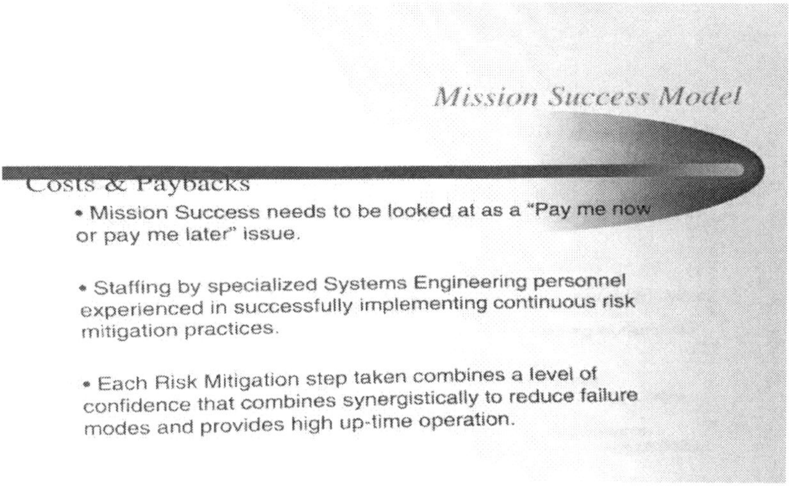

Exhibit MS – 26: MSM Costs and Paybacks or ROI.

Non-conformances in most USA companies range from 15 to 25%. Management may find this hard to believe since the so-called "Hidden Factory" seldom tracks this cost. As Exhibit MS – 26 says, Quality tradeoffs are very high risk and fall into the "pay me now or pay me later" category.

One story on-line at ASQ-TV Episode 2, "Culture of Quality" that caught my attention and worth noting here, was a story with Apple's iPhone 4 release. Rushed to market, it had dropout problems that consumers kept complaining about. This was caused when the phone was handled in certain positions of the holding hand (so called Death Grip)! After confirmation and re-design, the problem was solved and a recall was issued costing the company, some say in initially $125 million, but ended at the $525 Million dollars area! Definitely, "a pay me now or pay me later" real-world situation!

Chapter 11. MS Project Reviewing

NASA GSFC management reviews are part of the work world there. Remember in this high risk, technology-driven business communication keeping track of what is going on is imperative for success. At the functional Directorate level there are Weekly Meetings. At the Program Management level there usually are even daily…a heads-up we called it. With almost 100 different projects going on at one time close visibility on what's happening occupies most managers time. Before I retired, at the HQ level, new computer techniques were underway at HQ in the form of Management Dashboards that flowed down to the individual project task level to determine progress.

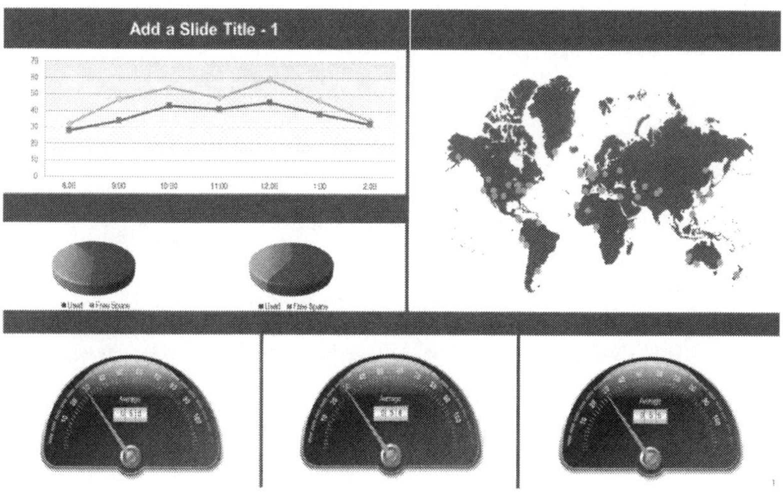

Exhibit MS – 27: Management **Dashboard Example.**

One can see the power of Management Dashboards (Exhibit MS – 27) for determining status of operations all over the planet in real time. This enables Managers at all levels to keep their fingers on the organizations pulse by not only transmitting data but information that can be used in daily decision making.

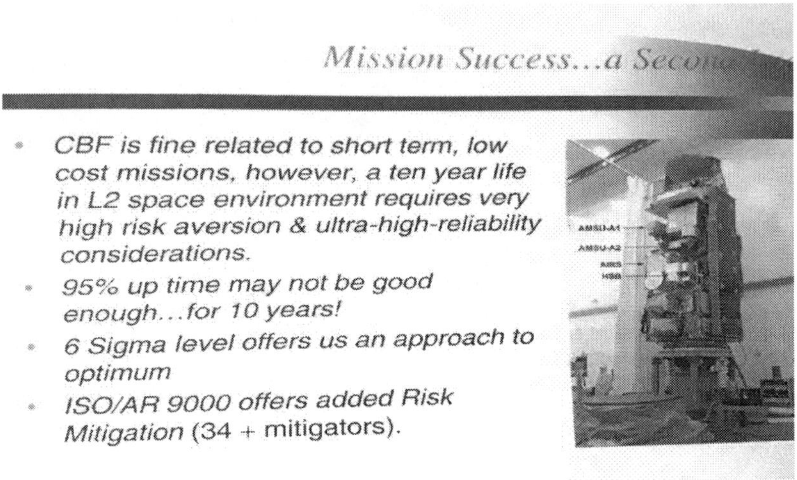

Exhibit MS – 28: MSN vs Cheaper, Better & Faster (CBF) Practices.

During the 1990's NASA HQ changed Administrative heads. The new one tried to change the culture by addressing a Cheaper, Better, Faster managerial philosophy. Many books were written about this new way of thinking in a high-risk business. At the Cape, where I travelled almost monthly during that time to oversee Team members, the saying among launch folks was amplified by a: cheaper, better, faster, splash retort!

Soon after another Administrator change this philosophy too passed. The book referenced below:

"Exploring the Solar System"

The abstract cited below, cuts to the core on its contents.

"But while NASA launched and achieved its goals for several missions developed under this "faster, better, cheaper" philosophy, five space science probes produced in this way failed before the decade's end. NASA soon thereafter backed away from this mode of planetary mission acquisition.

https://link.springer.com/chapter/10.1057/9781137273178_4

At this point in our mission success journey, I can't stop thinking of the story of the Russian General addressing his strategy vs the USA on our quest for Quality. According to him, he commented that you USA people concentrate on making your rockets perfect with Quality, we Russians feel differently. You hit us with one of your high-Quality missiles, we hit you with hundreds of ours.

Sometimes quantity beats quality! Perhaps he has a point, but one accurate shot is better than many wild ones that miss the target! The successful moon landing mission 50 years ago is a good example we can be proud of.

Chapter 12 MS Project Budgeting

Exhibit MS – 29 illustrates that Fed Gov Annual Budgets for NASA remained relatively stable during my years there. Annual figures were around $18 billion dollars. About less than 1% of the entire US Government budget. This is much lower than outlays for the Apollo Moon Landing Project that approached $26 billion but employed about 400,000 people in the then booming aerospace business. And the spinoffs were too numerous to mention here. From advances in medicine to scientific wonders like my watch, GPS, Weather, mapping, computers, Hi Def TVs, internet etc., etc.

From my viewpoint, no other agency in the Fed Gov has as much payback to the US citizenry as NASA.

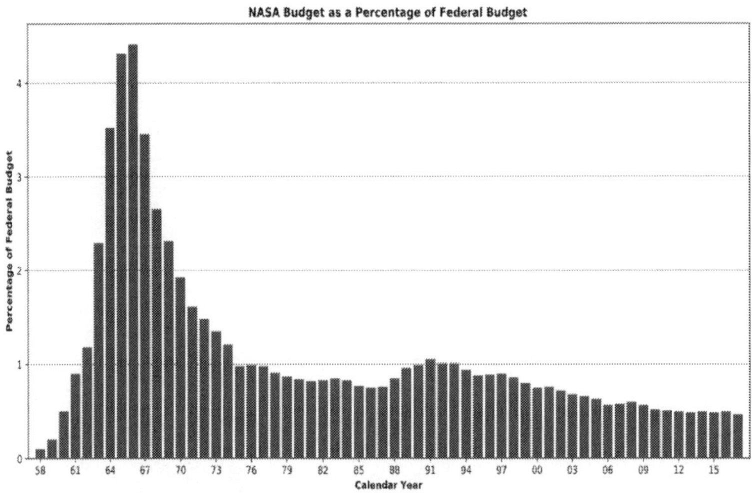

Exhibit MS -29: NASA Budget since inception as % of total Fed Budget.

(By 0x0077BE - Own work, CC0, https://commons.wikimedia.org/w/index.php?curid=35800864)

During my NASA tenure, we NASA, were confronted by a new commercial-driven philosophy that our work was too expensive, not worthy, always late and could be done cheaper. Thus, we all entered the cheaper, better, faster (CBF) philosophy for all new missions. Like many flavors of the month, no basis or scientific studies were made for these aspersions.

I delegate my thoughts on the matter to the experts on the MARS Failure Review Board of distinguished space management.

"The Mars Climate Orbiter mission was conducted under NASA's "Faster, Better, Cheaper" philosophy, developed in recent years to enhance innovation, productivity and cost-effectiveness of America's space program. The "Faster, Better, Cheaper" paradigm has successfully challenged project teams to infuse new technologies and processes that allow NASA to do more with less. The success of "Faster, Better, Cheaper" is tempered by the fact that some projects and programs have put too much emphasis on cost and schedule reduction (the "Faster" and "Cheaper" elements of the paradigm). ***At the same time, they have failed to instill sufficient rigor in risk management throughout the mission lifecycle. These actions have increased risk to an unacceptable level on these projects"***

(https://science.ksc.nasa.gov/mars/msp98/misc/MCO_MIB_Report.pdf)

Another Failure Review Board lesson learned is worth noting on the IMAGE Satellite.

https://image.gsfc.nasa.gov/publication/document/IMAGE_FRB_Final_Report.pdf

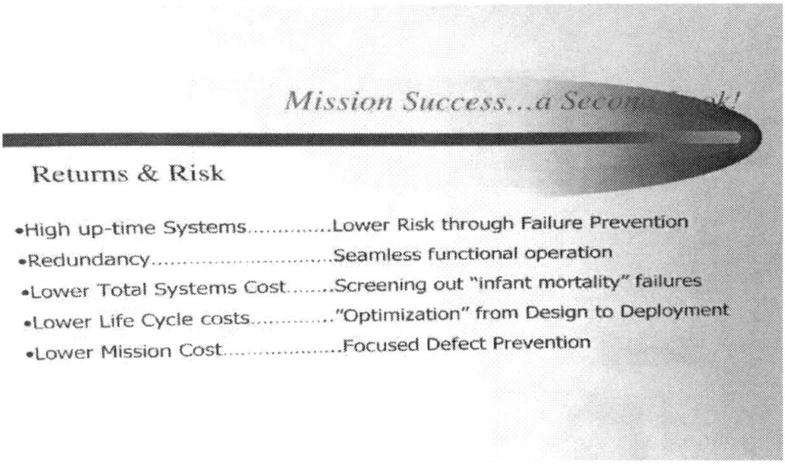

Exhibit MS – 30: Mission Success: A Second Look Chart

Mission Success can be looked at as a second look insurance policy. It offers a high up-time payback, additional redundancy, lower total systems cost lower life cycle costs and lower mission costs as illustrated in Exhibit MS – 30.

Costs** - Always a major consideration for every new mission, are carefully monitored and assessed by management oversight.

Costs associated with the establishment and operation of a ground segment are highly variable and depend on accounting methods. According to a study by Delft University of Technology, the ground segment contributes approximately 5% to the total cost of a space system. According to a report by the RAND Corporation on NASA small spacecraft missions, operation costs alone contribute 8% to the lifetime cost of a typical mission, with integration and testing making up a further 3.2%, ground facilities 2.6%, and ground systems engineering 1.1%.

Ground segment cost drivers include requirements placed on facilities, hardware, software, network connectivity, security, and staffing. Ground station costs in particular depend largely on the required transmission power, RF band(s), and the suitability of preexisting facilities.

Control centers may be highly automated as a means of controlling staffing costs.

**Based on a model described in *Space Mission Analysis and Design*, third edition, by James W. Wertz and Wiley J. Larson

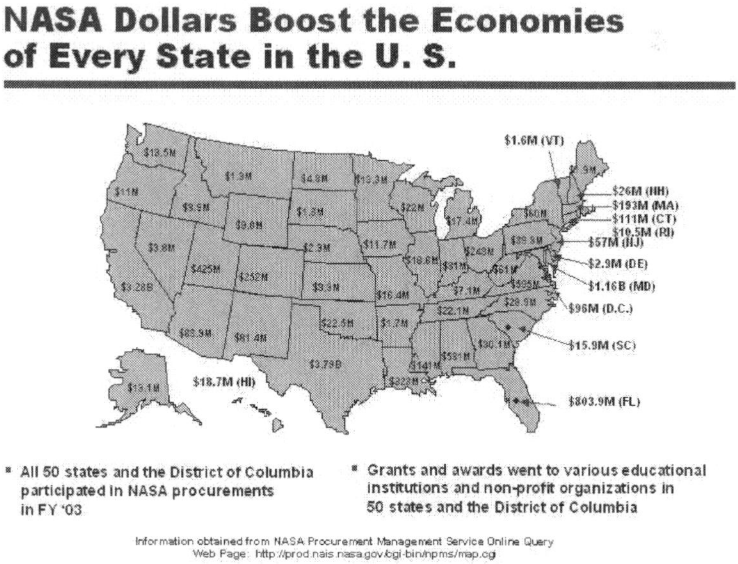

Exhibit MS – 31: NASA Dollars by USA States

(By NASA Acquisition Internet Service - http://prod.nais.nasa.gov/cgi-bin/npdv/map.cgi, Public Domain, https://commons.wikimedia.org/w/index.php?curid=4292115).

Exhibit MS – 31 illustrates on a USA map, the distribution of NASA Dollars over the entire nation. High technology areas and NASA Center locations provide significant long-term employment for the past 50+ years. In return, all technology we now take it for granted such as GPS, Weather, International Space Station and hundreds of scientific satellites are maintained seamlessly for 24/7/365 operation.

Chapter 13 MS Lessons Learned

Exhibit MS – 32: 50 Years of Mission Assurance at NASA

We start here with my 50 Years of Mission Assurance pitch (See Exhibit MS – 32 Title page) I gave years ago at the NASA GSFC Annual Contractors Day Event. Hundreds of Lessons Learned are now in the NASA HQ data base already for the new PM organization to review for knowledge of what to do and more importantly what not to do for achieving successful space missions.

Annually. NASA GSFC has a Contractor's Day Conference Event to cover recent industry findings for all interested companies and governmental agencies. We were approaching the 50th Anniversary at NASA (founded in 1958) and wanted to present our findings on analyzing over 4,000 on-orbit anomalies recorded over the past several years. In addition, I discovered a new analysis tool that I used to show how to prevent and mitigate future anomalies in spacecraft missions. We cover it here in our discussion of Barrier Analysis.

In addition to my presentation, we had a new one on counterfeit parts and the space technology industry impact of such a growing menace. A most difficult potential failure to screen out, the audience was given several methods to catch these culprits before using them in space projects. Even doing destructive testing on lots of devices and de-lidding them to assure part identity of legitimate supply was covered. Warnings of dealing with factory made supply not distributor or second of third supply sources was a big prevention measure. This is becoming more difficult as many parts are now made on foreign countries because of cheaper labor and lack of USA manufacturing plants.

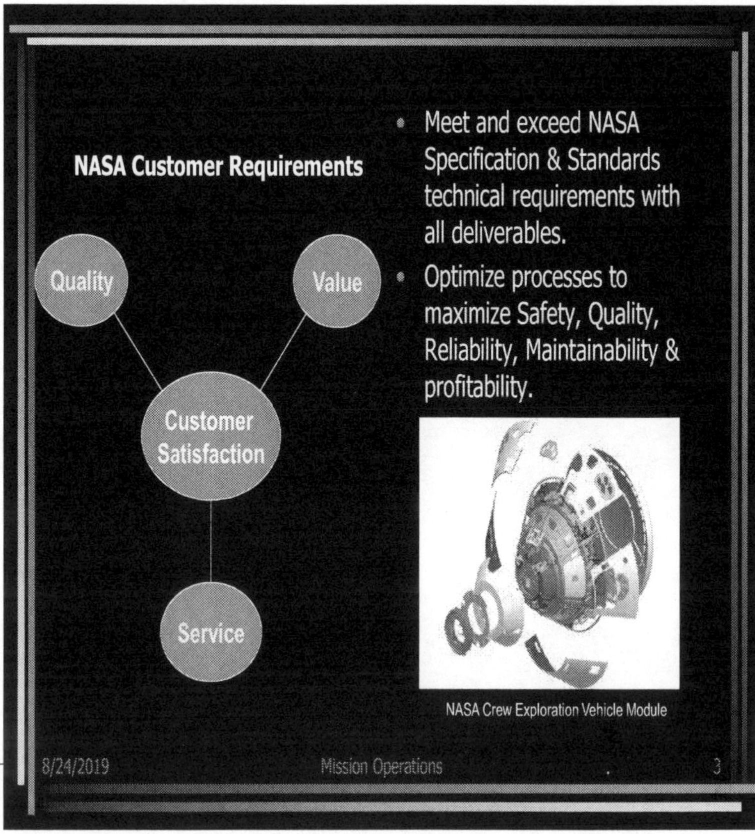

As shown in the above Exhibit MS - 33, NASA Quality, Value, Service (QVS) focus is on Customer Satisfaction. This is done by meeting and exceeding Requirements, Specifications & Standards. And by optimizing Safety, Quality, Reliability, Maintainability and Cost of Quality.

What's Driving NASA GSFC Anomalies

Causes	Effects	Possible Ramifications		Preventative Actions
1. Design	Heritage, Similarity	Early life failures	Mission impact	Pre-Award Surveys, Assessments Peer & Mission Assurance Reviews
2. Workmanship	Lack of Standards	Anomalies	Quality escapes	Quality Plans & Procedures, AS9100 GSFC-STD-1000, 1, CMMI, AIAA
3. Processes	Deviations/ Waivers	Continual faults	More escapes	Mission Plans, Procedures & Work Instructions
4. Documentation	Unclear requirements	Mission impacted	Difficulty to remedy on-orbit	CM, CM, CM...PRD, CDR, Final Delivery Verification
5. Schedule	Short cutting	Waivers	Partial successes	Provide Contingencies....Plan A, B, C...!
6. Parts	Weak sisters	Shortened life	Catastrophic consequences	Selection, Qualification, Screening, Inspection & Testing
7. Environmental	SEUs	Mission Susceptibility	Lost mission time	Do EMI/RFI Testing! Design in RFI protection!
8. Software	Mission losses	Inability to fix	Costly S/W Patch workarounds	CM on all changes! CMMI, Validate & Verify Testing process, use Hi-Fidelity Simulators
9. Testing	Circuit failures	Premature failures	Lost science	Incoming, In-process, Final, RAT/RIDT, End-to-End, Full-up testing, ATP/Env. TP
10. Operator	Induced failures	Missed Science	Lost science/missions	Certification & Training, Re-certification, Poor typing...757 Columbia....R...instead of ROZO!
11. Ground Systems	Weak link in the chain	Multiple hits	Multiple mission losses	Backup Plans, Software validation, Patch testing

8/24/2019 Mission Operations 6

MS – 34: Anomaly Drivers on NASA GSFC missions.

In reporting on Mission Operations space craft on-orbit performance, we summarized our findings of over 4,000 spacecraft and tracking station anomalies in Exhibit MS – 34 entitled “What’s driving NASA GSFC Anomalies.

We broke down anomaly causes into 11 Cause Categories as shown above. Their effects were then analyzed, and possible ramifications noted. Each cause was then associated with our recommended preventative actions that for future missions provided an invaluable beginning to future risk management studies.

For further information on the NASA Lessons Learned Data base refer to the Website listed below.

(Https://www.nasa.gov/offices/oce/functions/lessons/index.html)

Chapter 14. Mission Success Model

As seen in Chapter 1, Exhibit 1, where our subject began, we see our MSM model taking form. A Mission Success road map follows and starts at the bottom with Project Lifecycle. This path takes us through the Design, Fabrication, Integration & Test, Spacecraft Integration and Launch stages of building a spacecraft.

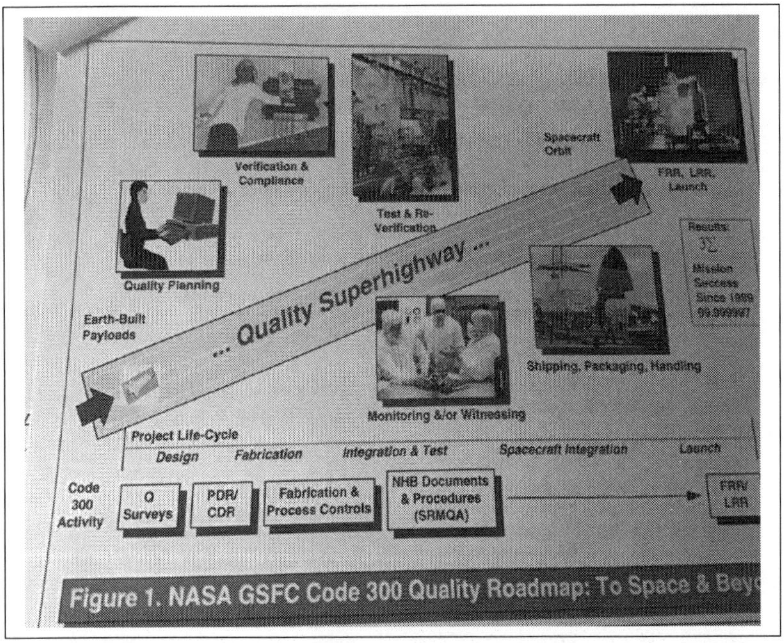

Exhibit MS - 35: NASA Quality Roadmap Process Flow Chart

NASA GSFC Code 300 Activities (Mission Assurance) starts with Surveys, continues with Design Reviews, Fabrication/Process Controls, utilization of NASA Documents/Procedures finally ending up at Flight and Launch Readiness Reviews. All the bases are covered by this roadmap designed to meet six-sigma levels of results.

However, this goal needs additional 21st Century techniques such as what we have described plus the use of computer driven programs that act as cost effective tools to simulate design performance.

Although all the policies and procedures assist in the control effort, there is nothing today that can replace the human interface or trained, experienced technical personnel it takes to oversee activities all along the way.

Exhibit MS – 35 lists the detailed tasks to be performed by Project personnel. We have used it as a basis for verifying completion of the mission success roadmap. One can even score a rating of implementation for continuous follow-up during the entire program life cycle.

NASA GSFC Supplier Chain Workshop 2007

Exercise: What should the future systems BAQ score be?

Barrier Analysis Quotient (BAQ)

Barrier Level	Barrier Elements to Measure	Range of Scores (Min. – Max)	Assessment Score
A. Big Three	1. Policy	1 to 10	
	2. Procedures	1 to 10	
	3. Surveillance	1 to 10	
B. Top Eight	1. AS 9000	1 to 10	
	2. Training	1 to 10	
	3. Certification	1 to 10	
	4. Requirements	1 to 10	
	5. Support	1 to 10	
	6. Standards	1 to 10	
	7. Testing	1 to 10	
	8. RE-Views	1 to 10	
C. Mission Success + 8	1. Reliability	1 to 10	
	2. Benchmarking	1 to 10	
	3. Driver Metrics	1 to 10	
	4. Success Checklists	1 to 10	
	5. Dashboards	1 to 10	
	6. Pareto	1 to 10	
	7. Mapping	1 to 10	
	8. Six-Sigma	1 to 10	
	19 Elements Total	19 to 190	
		BAQ	

1/16/2020 Mission Operations 30

Exhibit MS – 36: Barrier Analysis Assessment Table

Let's now apply the Barrier Analysis tool to our Mission Success Model. The Exhibit MS – 36, puts it in perspective. I created a Barrier Analysis Assessment Table to determine Barrier element scores for baselining requirements for compliance determination.

Exhibit MS – 37 reveals anomaly mitigation and interdiction measures. To prevent failures, we already indicate the actions to prevent anomalies. They are identified as "A" anomaly sources. Next we see our classic anomaly barriers identified as "B" classical countermeasures is the term I use. All barriers are focused on the target at risk "C" either the Spacecraft itself, the Ground System (including the launch site, pad and missile). Not to forget the vast communications networks from tracking stations to Mission Control to Mission Operations…a lot of bases to cover!

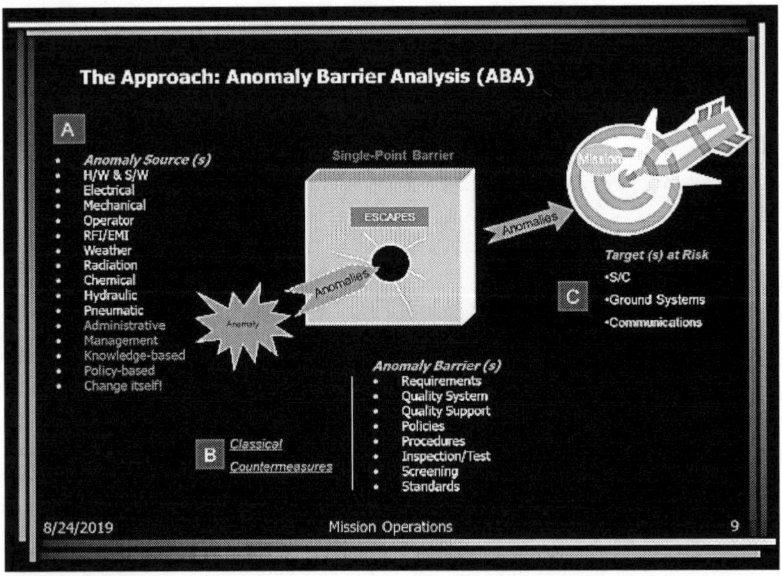

Exhibit MS – 37: Barrier Analysis chart for anomaly mitigation

Before we continue to explain the power of Barrier Analysis, let's look at the anatomy of a failure. In most cases as the Exhibit MS – 37 illustrates it shows only one obstacle for an anomaly to become a mission failure target. Often time in Catastrophic incidents it takes more than one obstacle to cause disaster. Perhaps it takes two or three conditions, all in phase, to be overcome to prevent failure. See all anomaly sources involved. 15 of them are shown here (A). Next 8 barriers need to be overcome (B). finally, on spacecraft launches and missions at least 4 targets (C) are at risk: Spacecraft, Ground Systems, communications and, in addition, the Launch vehicle itself.

A good example of this several in-phase conditions failure was recently shown on the TV show "Air Disasters" was with the Boeing 737 – 3Q8 owned by Garuda Indonesia Flight 421. On January 16, 2002, the flight encountered severe thunderstorm activity during an approach to its destination, suffered a flame out in both engines, and crash landed in a shallow SOLO River near Surakarta, Indonesia, resulting in one fatality and several injuries.

Post NTSC Investigation Report showed three conditions that were the cause: one jet engine failure due to out of specification conditions resulting from the local rain/hailstorm. Engine restart was lost also due to low on-board batteries (24 volts down to 22 volts) and finally on-board radar shadowing caused a clear landing path when there was none. Fortunately, the pilot did a Sully Sullivan on the Hudson type landing that prevented total loss of life.

Thus, the reason for more barriers to prevent failure escapes such as Flight 421 demonstrated. Several corrective actions on engine starting and radar readings were implemented, however, as the report showed flameouts are a concern during heavy rain/hailstorms. In those cases, however battery power saved the day on these flights.

Next let us look at all the challenges. I offer this from a Reliability Symposium Presentation I attended where USAF Lt. Col. Susan Hastings, NRO/DSIE, so nicely made up (Exhibit MS - 38 below) that showed mission success turnaround in spite of increasing space systems and vehicle complexity and push to reduce costs in the acquisition practices and rise of consumer electronics. She advocated even additional surveillance for the verification process.

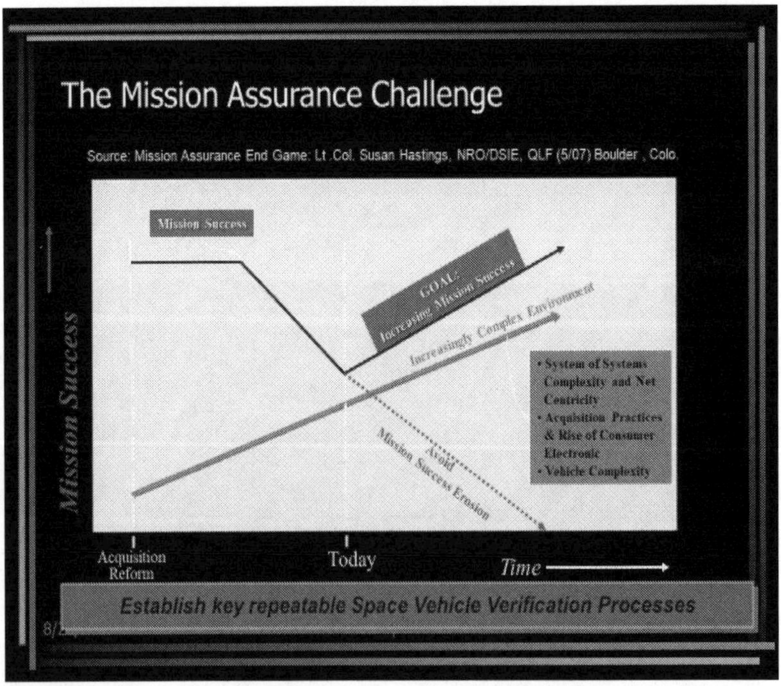

Exhibit MS – 38: Mission Success as a Mission Assurance Challenge.

And now back to Barrier Analysis, Exhibit MS – 39 Applying Multiple Quality Barriers (Rx) that not only one layer of anomaly barriers is required to achieve superior anomaly interdiction, but a second level is necessary. These are present Aerospace basic and improved Risk Management measures, but a third level is required as laid out in this book.

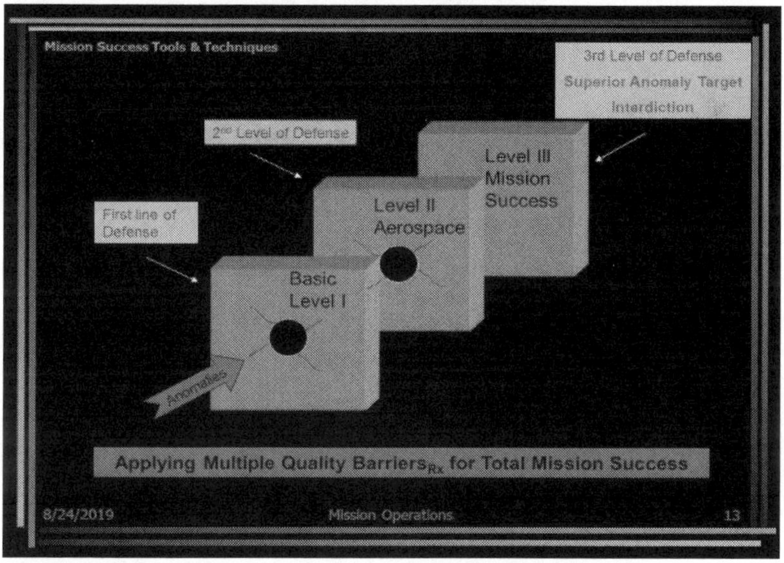

Exhibit MS – 39: Applying multiple barriers for Mission Success

Next Exhibit MS – 40 below still shows gaps present in our model. These escape areas allow oftentimes random failures that cause mission failure when one least expects it…like Apollo 1 an unthought of catastrophic disaster that may have been prevented with worst case scenario risk analysis.

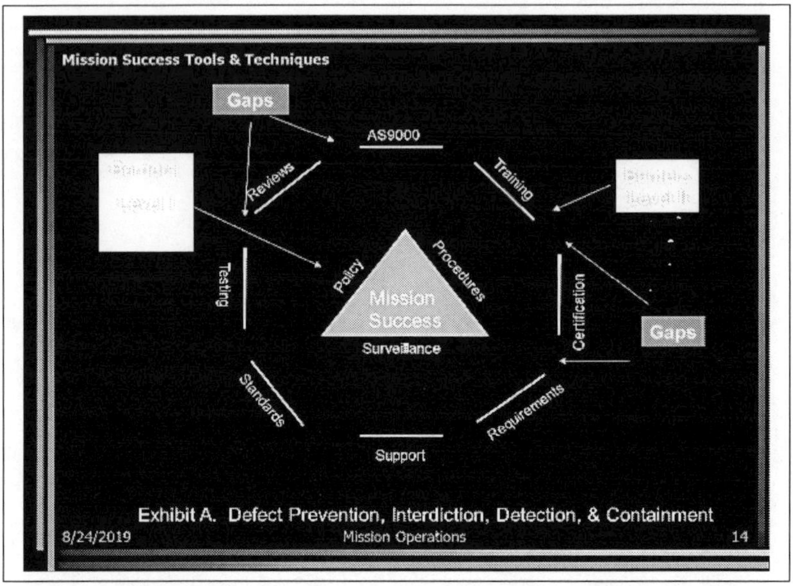

Exhibit MS – 40: Level I & II Barriers for Mission Success

This results in closing in the anomaly gaps that remain This results in closing in the anomaly gaps that and as Murphy's Law states, they happen when least expected and sometimes resulting in catastrophic failure.

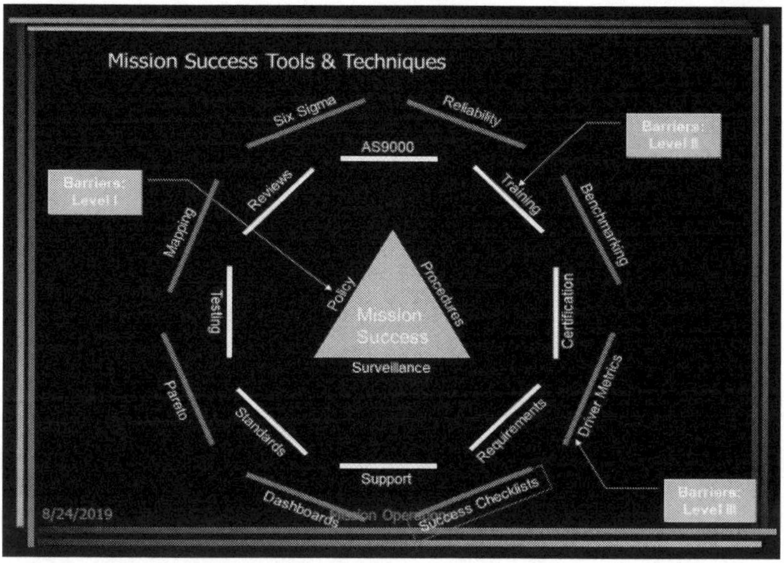

Exhibit MS – 41: Third level of barriers for achieving Mission Success.

Exhibit MS – 41 shows Barrier level three, that applied fills in the anomaly gaps. Such additional measures like the use of Success Checklists, Pareto Analysis, Mapping, Benchmarking, Reliability predictions and ECAP Analysis and 6 Sigma techniques.

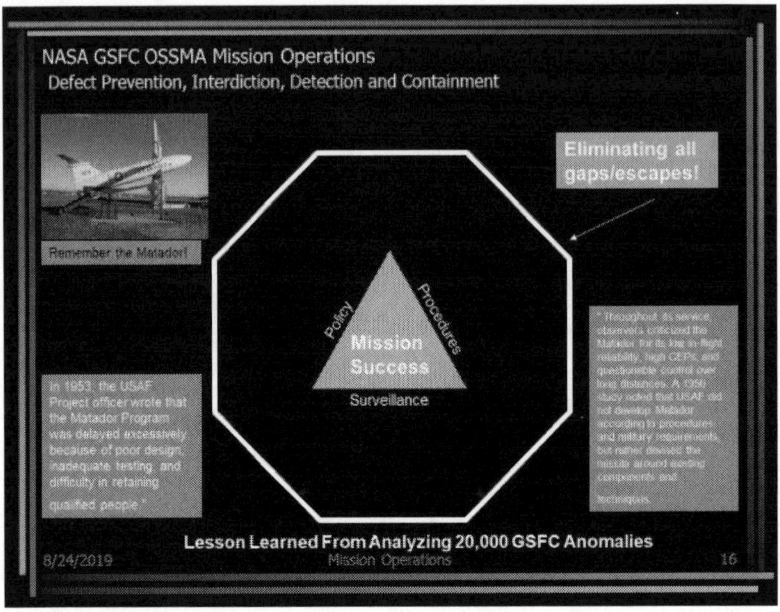

Exhibit MS – 42: Mission Operations Lessons Learned Anomaly Analysis

Exhibit MS – 42 shows the Mission Success Three-Legged Stool made up of Policy, Procedures and Surveillance when combined eliminates all anomaly gaps and escapes.

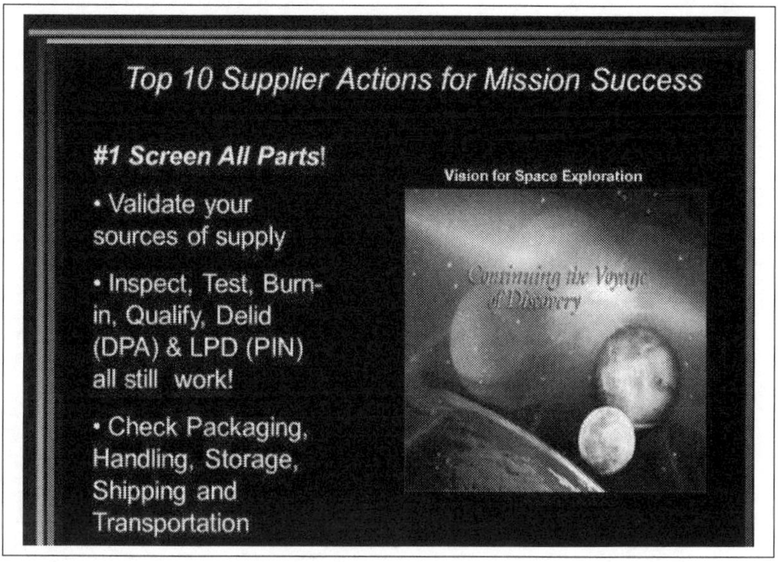

Exhibit MS – 43: Supplier Actions for Mission Success

Exhibit MS – 43 gives the Number One Action for supplier and NASA to perform for Mission Success that is to screen all parts used in the spacecraft mission. This should be done after validation of supply sources is performed. Then microelectronic devices especially need further inspection, burn-in to eliminate early life failures and DPA even PIN testing. As I recollect from the Rockwell B1B Test Pilot, the biggest fear he had was loose particle or FOD inside the plane. Such things as coke bottles and adjustable wrenches can ruin your day when you are doing flight qualification testing. The same holds true with microelectronics devices…short circuits suddenly appear often causing catastrophic failures. ESD is also a big headache with these devices, therefore packaging, handling and transportation are critical steps in the supply chain.

Contractor requirements are a key anomaly mitigator. We have provided a Top 10 List of actions they need to perform for helping achieve mission success. First Screen all parts to weed out marginal components. Exhibit MS – 44 goes into detail on the how to do and what screening is required.

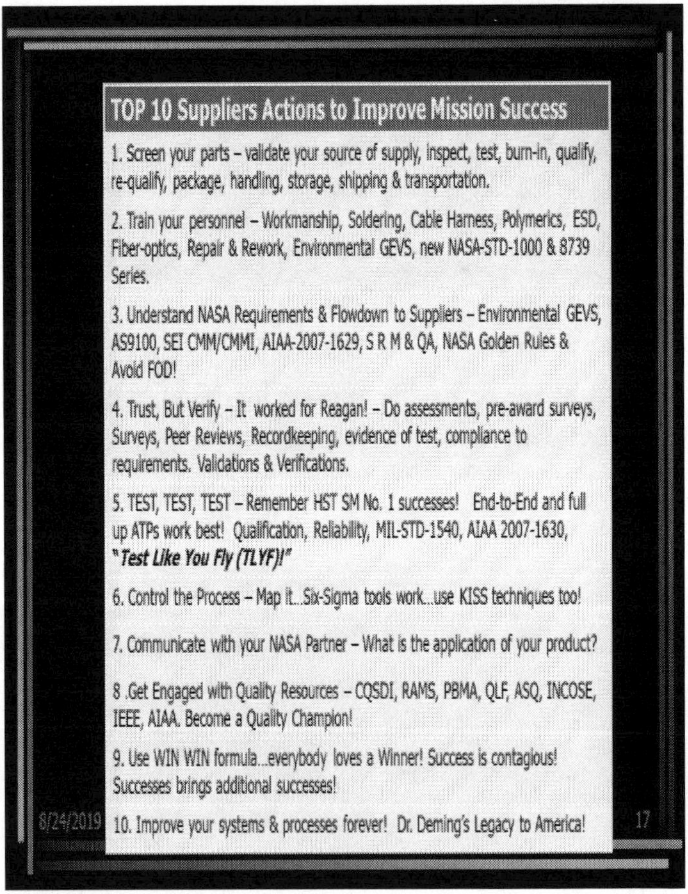

Exhibit MS – 44: Top 10 Suppliers Actions to improve Mission Success.

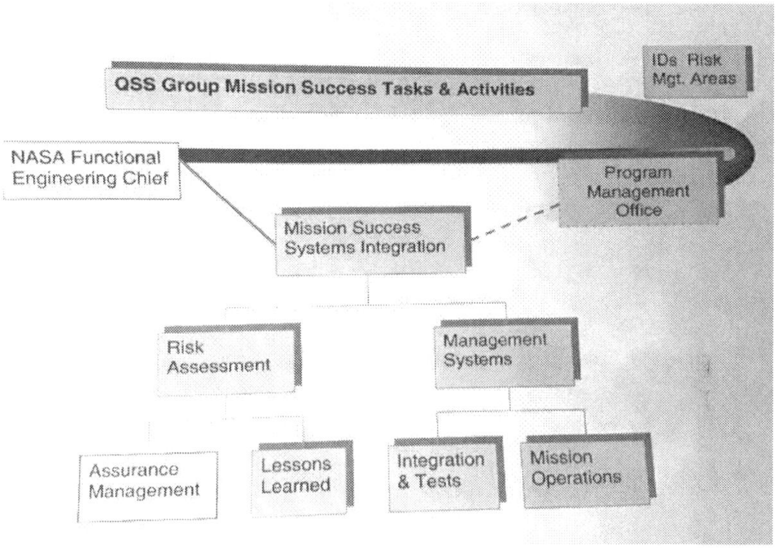

Exhibit MS – 45 MS Tasks & Activities

As one sees in Exhibit MS – 45, Tasks range from the NASA Engineering Chief whose job is to assure compliance with NASA PMO compliance to requirements to the PMO who oversees all operations associated with the mission. This delegated authority flows and crosscuts across all operational areas with Assurance Management conducting their assignments including Risk assessments and applying Lessons Learned, witnessing Integration & Testing activities as well as Mission Operations.

So, what are we trying to avoid with this new focus on Mission Success?

Of course, mission failures are the obvious answer! But the other consideration is what we called at NASA anomaly. These are costly, time consuming and often a nuisance to all involved.

Just what is an anomaly? Simply defined, an anomaly is the unexpected behavior of a system. Anomalies represent an undesirable condition event and are usually costly. First cost is tracking, followed by analysis and thirdly corrective action. Not to mention loss of science data or information. And of course, reporting, classifying and further tracking and opportunity cost of spending time and money on doing other things. Anomalies are indeed costly!

Some of my time at NASA Mission Operations was reporting anomalies. At the time we had over 40 birds (our term for satellites) to report daily on. Well over 4,000 were recorded and reported. We received inputs from the many NASA Tracking Stations all over the planet and every hick-up was recorded. We immediately alerted management on any serious finding and reported weekly on all observations. Our job was preventative on nature. To analyze trends and take appropriate action, both preventative and corrective. Annually, we contributed to the Center Director's OAGS Report.

A big item we reported on weekly was termed "Operator Error". Remember one of the most important part of any space system is the ground systems involved to communicate with space missions. Just the EOS Program, with three satellites: AQUA, AURA and TERRA had terabytes of data streaming daily from their scientific studies.

I worked on verifying the NASA GSRD (Ground Systems Requirements Document) prior to EOS Aqua Satellite launch and hundreds of points in the vast network were checked and re-checked. I recall the saying, that was denied by Bill Gates, that he said that 640K of memory is more memory than anyone would need! And now we talk terabytes as normal! In fact, this new HP Laptop, with lighted keyboard, finally, has 1 Terabyte of storage!

David Harland and Ralph Lorenz' book ***"Space Systems Failures"***, provides valuable insight into mission failures and summarizes both types of failures and specific functional systems failure origins also.

Design and Environment made up 46 %, followed by Parts/Quality 24%, undetermined another 20%. Using Pareto's Law of Distribution, the significant 80/20 Rule homes in on the "Failure Drivers".

Spacecraft-wise, failures involved 50% of Spacecraft Telemetry, Command & Control, Guidance & Navigation and Power. Payload-wise, nearly 30% involved Sensors & Communications.

In Summary, 70% on-orbit failures involved Spacecraft and 30% Payload, nice Pareto distribution. So, again Spacecraft reliability is extremely important, if you can't get payloads into orbit, your efforts are "down the tubes" as they say.

So, what to do? You might be thinking? Well Dave and Ralph focus on taking the following actions: robust/careful design, wider margins, redundancy, expect the unexpected interactions in sub-systems, simulation and finally, Testing (we called it "Test as you fly, Fly as you test"!).

Exhibit MS – 46 is an overview of Jet Propulsion Laboratory model of screening out defects. Using mission failure modes risk elements at the top it flows down to Rules, material selection and robust design. At the same time analysis leads to qualification, testing…life, simulation, assembly, and finally, system level performance testing. All lead to their mission success criteria. I obtained this graphic while attending annual Reliability Assurance Management Symposium training where JPL was always last but not least when they always gave an outstanding performance that leads to a standing ovation. I have high respect and regards for JPL's second to none scientific staff.

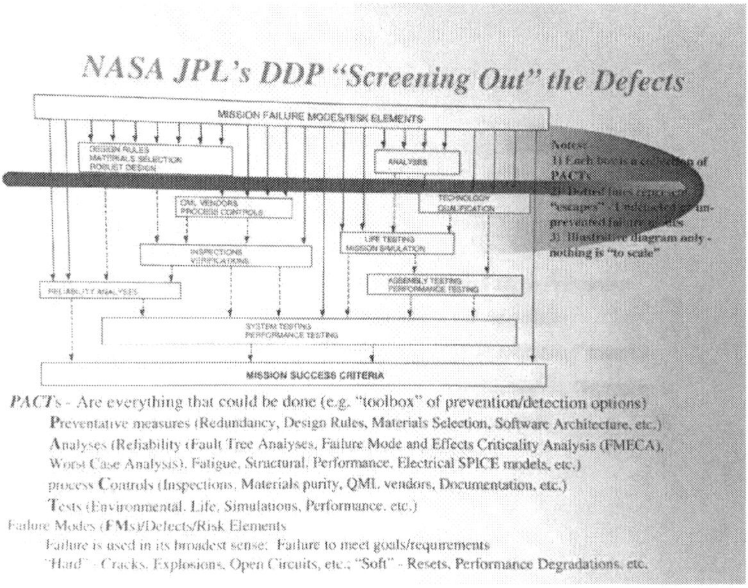

Exhibit MS – 46: NASA JPL Mission Success Criteria Graphic.*

*Courtesy of "RAM Symposium Las Vegas 2008"!

Chapter 15. Mission Success - Success Factors

As often wished for by a failed mission Program Manager "why didn't they tell me what to do?" for hindsight is 20/20. Foresight too is 20/20, if you have a map on how to get there! Cutting corners to meet schedules and failing is probably the chief contributor for mishaps.

Airplane accidents often happen with the most experienced pilots. Just look at one that happened when I worked on the B1B at Martin Marietta. The Rockwell veteran test pilot went down on a low terrain test flight in the dessert because he forgot to shift the planes fuel load prior to executing the maneuver.

Accidents happen and more so in high-risk applications as the Space Program or Aerospace, in general. ***"An ounce of prevention is worth a pound of cure"*** was said by Ben Franklin in 1736. This led to the first Union Fire Company in Philadelphia, PA.

As we have seen in this book accidents happen, and mission success happens also and as Franklin so wisely said one must provide for the worse-case scenario and with proper planning and oversight the best-case scenario will apply.

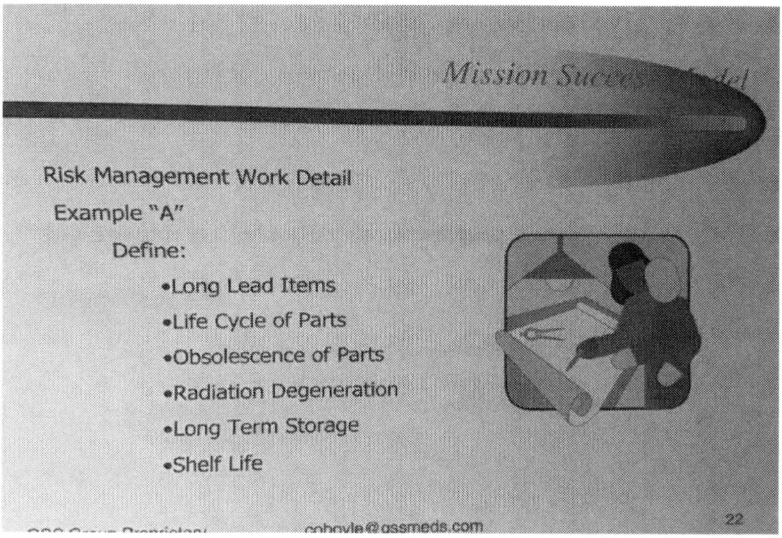

Exhibit MS – 47: MS Model Risk Management Detail.

I am amazed by this time in the 21st Century how little the main driver for success is in our historical space endeavors. As John Glenn once said, "my only problem was that this contract to build this space vehicle went to the lowest bidder!" Well, John also knew that all qualified bidders must meet the space RFP requirements in that, they all had mature Quality Assurance Departments that was responsible for work oversight from design to build stages of manufacturing.

As illustrated in Exhibit MS – 47, Risk Management details focuses on space craft parts criticality such as shelf life, Radiation susceptibility, long term storage, obsolescence, lead times.

NASA has a preferred parts list (PPL) that is recommended for space applications and is constantly updated by Space Center Parts Engineering personnel.

I entered the civilian high technology business in the summer of 1961 and at that time a formal Quality Control organization existed with a complete department whose responsibility ranged from Clausing the purchase order requests with standard Quality standards such as Mil-Q-9858 (April 9 , 1958, Revised December 16, 1963 to Mil-Q-9858A.

And dropped in 1996, in place of ISO 9000! And who the hell ever did that should be jailed!

For right after, the Europeans copied our Mil-Q-9858 as a way of stopping US trade goods into their countries and sold it back to us later as ISO 9000! How about that one folks??

Thank heaven for those West coast airplane manufacturers who went beyond ISO 9000 because it was not good enough for their industry and added 34 more requirements in a new document called AS9000, finally adopted many years later by NASA at the turn of the Century after my pleading and begging. "No budget for that Chuck", was their reply. Then came Columbia and Hubble. The lesson of Challenger still had not changed the culture. That was described so eloquently as done by Dianne Vaughan entitled, ***"The Challenger Launch Decision"*** that the cause was a "Normalization of Deviance!" Groupthink was a pattern in US policy thinking and she used examples of Bay of Pigs, Pearl Harbor and Viet Nam as similar examples.

We start at reviewing the past anomalies and failures to forecast what measures, controls, requirements and verification/validation efforts are worth pursuing in our efforts of achieving the goal of "**zero defects**"!

Well, as the chapters we have read shows many variables are at play in space missions. And always costs are primary considerations as was seen in all the USA Space history. We saw where just one shuttle disaster caused billions of dollars and loss of 16 Astronauts that are "priceless"! A second accident was certainly a factor to not proceed with more shuttle-based space missions.

And a jaded public did not help our space efforts. At the end launches were just another day at the office and even the news media reduced coverage of the big missions.

Technology today can bring us more "bang for the buck"! Semiconductor design is more reliable, less parts are the result. Recoverable Boosters are a reality with them flying down like a big bird back to the launch pad, ready for the next mission. Fed Gov support still important in a still High-Risk business as space flight still is!

Exhibit MS – 47A: List of Mission Success Factors in Space Applications

Quality (QMS)	AS9000	Inspection	Testing	Qualification
Reliability	Redundancy	Modeling	RAT/ECAP /TRAC	RIDT
Risk Mgt.	Assessment	Predictions	Mitigation Models	PRA Package
Safety	Hardware	Software	Worse case Analysis	Single Point Failure
Systems Eng.	SEMP	Requirements	Full up Testing	End to End Testing
Software	Software Plan	Verification	Validation	Avoid Heritage

As listed in Exhibit MS – 47A above, each functional organization (top down) has responsibility for applying their requirements to the specific spacecraft hardware/software both in-house and out-of-house workplaces. Verification is continuous throughout the project lifetime through the launch and on-orbit Certification.

Exhibit MS – 48 covers the impact on Mission Assurance Requirements (MAR) in detail showing deliverables and risk mitigation focus.

Exhibit MS – 48: Mission Assurance Requirements (MAR)

Contents	Deliverables	Mitigation Focus
1. General	SSMA Plan	Quality Organization
2. Quality Management System	QMS	Process Quality Control
3. Systems Safety	SSM Plan	Risk Management
4. Reliability	RP Plan	Fault Tree Analysis
5. Software Assurance	SA Plan	V & V Efforts
6. Workmanship	Training/Certifications	Soldering, CC, R&R, ESD Certification
7. EEE Parts	PC Plan	Parts Control Board (PCB)
8. Materials & Processes	MIP Plan	Non-standard Control/MRB
9. Conformal Coat & FOD Control	Training/Certifications	Failure Prevention

10. Metrology & Calibration	Recall Control	Standards compliance
11. GIDEP & Problem Advisories	Compliance	Industry Alerts
12. End Item	End Item Data Package	Verification
13. Acceptance Data Packages	Contract requirement	Acceptance Package
14. Compliance Matrix	For review by NASA	Verification
15, Quality Survey	Approved Supplier	In-plant Audits/Residence

All spacecraft hardware suppliers are required to meet NASA GSFC Mission Assurance Requirements (MAR). As standard practice on new GSFC missions, a Spacecraft MAR document is prepared, using the 15 requirements, that is tailored to the project. Many times, I myself was part of a NASA GSFC Survey team that conducted in-depth audits of supplier's capabilities and facilities prior to any order being awarded. Even after award, audits were periodically performed. In-Plant residents had checklists to maintain of their conducting conformance audits of their sites. Unlike today's FAA practices at Boeing, no delegation was given in a "trust, but verify situation", as Dave Packard so nicely enforced as Deputy Under Secretary of Defense. As today's saying goes "**verify, then trust**" is the industry byword.

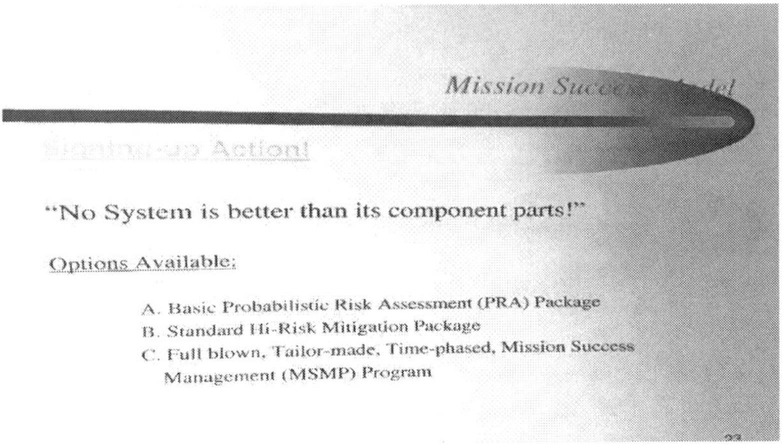

Exhibit MS – 49: Conclusion of MS Model

This is where we ended our pitch as our Apollo Era NASA Parts Engineer Joe Osche, so nicely said, "*No system is better than its component parts*!" Management either takes path of uninsurable uncertainty of mission success or uses the proven measures and requirements of success discussed herein. There is no easy way or cheap way to deal with the high-risk of designing and building Aerospace products.

Lots of improvement work still exists to refine the present 21st Century Quality model to meet and exceed present and future spacecraft specification and mission requirements.

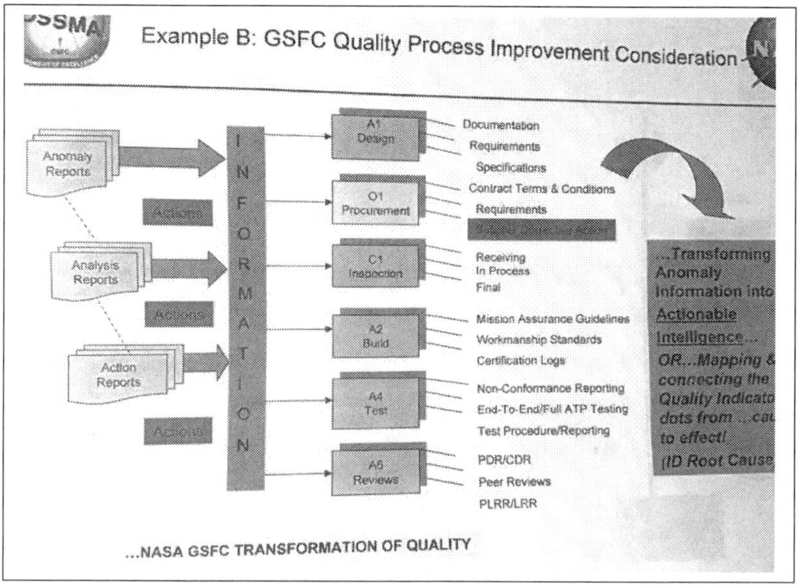

Exhibit MS – 49A Transformation of Quality Process Improvement

As Exhibit MS – 49A illustrated a transformation through Quality Improvement can utilize anomaly analysis information to all stages in the Spacecraft life cycle where corrective actions are embedded into existing requirements resulting in elimination of defects e.g. Zero Defects!

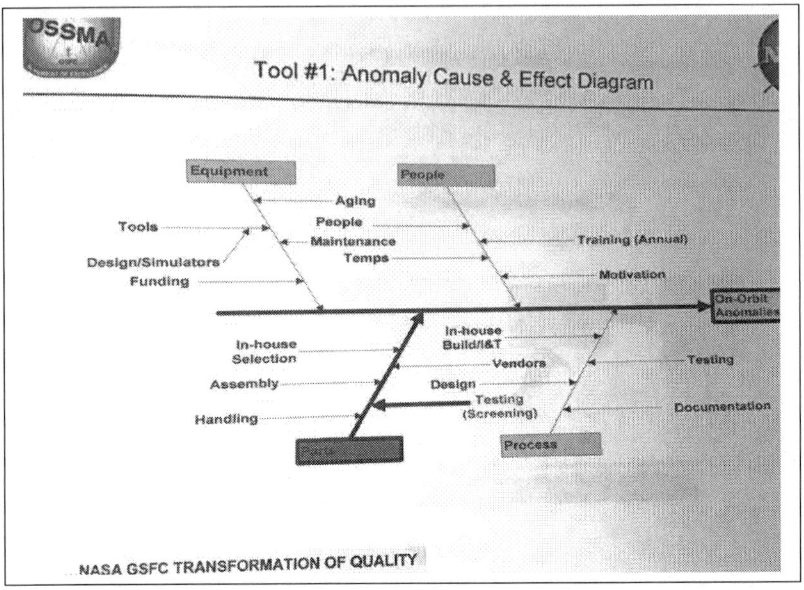

Exhibit MS – 49B: Fish Diagram Tool for Transforming Quality

Exhibit MS – 49B breaks down anomaly causes into four categories:

1. **Equipment** – Calibrated and maintained with records.
2. **People** – Trained/Certified annually.
3. **Parts** – Preferred, Qualified, Tested, avoid heritage/counterfeit!
4. **Processes** – Certified with records.

Together, each provides a new map for implementing corrective action for root causes of anomalies, incidents and failures. Note the focus on main drivers in each category and heavy emphasis on the Parts bone on the diagram. This analysis is all part of the on-going quality improvement process that must go on to provide future mission success.

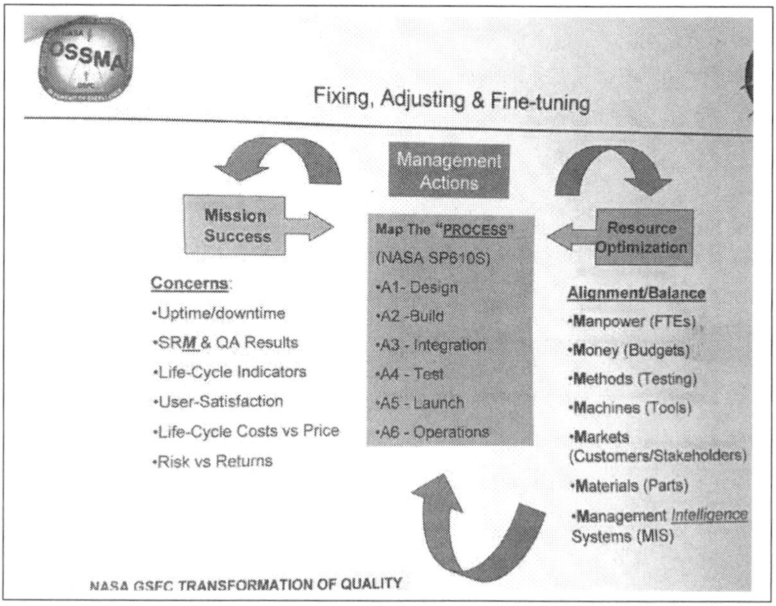

Exhibit MS – 49C: Management Actions for Quality Transformation

Note the emphasis in Exhibit MS – 49C on fixing, adjusting and fine-tuning the Process (A1 – A6). Mission Success concerns and resource optimization alignments/balances are constantly at play in the process. The successful project is one in which each ingredient in the Mission Success recipe is efficiently and effectively utilized by the management team.

Besides Mission Success described here, NASA and all aerospace companies must continue to improve the system forever as Dr. Deming predicted.

Four big challenges are listed for management to be focused on:

1.Continuous Process Improvement is necessary since just benchmarking the industry in year 2020 shows many voids from being implemented to close the fault escapes and gaps that still exist in quality management systems today.

2. Cost of Quality (COQ) models are rare to find. Attacking the four basic Prevention, Appraisal, Internal & External Costs need to be managed.

3. Quality improvement programs are also difficult to find, except in Japan with their Phil Crosby Zero Defects programs. Use of a Program “Help Desk” was a recommendation year ago but still not practiced by programs.

4. Software activities are in the dark ages of control. Just ask Boeing! Software Assurance requirements need to be developed and implemented just like hardware.

5. Lastly, statistical tools like Statistical Process Controls (SPC) are rarely seen in auditing Quality Programs.

If you think your Prevention and Appraisal Costs are in order look at the Checklist below and compare tools that are used with what is available!

Action Required: Management needs to transform Mission Success from a cultural trait to a program management requirement (See Exhibit MS – 49D).

Table MS – 49D: Quality Management Systems Tools

Assurance Tool	Evidence (X)	Corrective Action Required
Process Flow Charts		
Process Run Charts		
Pareto Chart		
Checklists		
Check Sheets		
Histograms		
Fishbone Diagrams		NASA FRB use in many reports
Benchmarking		
Scatter Diagrams		
Control Charts		

Chapter 16. Mission Success Application (JWST)

For your reference - NASA James Webb Space Telescope (JWST) Website: https://www.jwst.nasa.gov/

The James Webb Space Telescope, also called Webb or JWST, is a large, space-based observatory, optimized for infrared wavelengths, which will complement and extend the discoveries of the Hubble Space Telescope. It launches in 2021.

It will cover longer wavelengths of light than Hubble and will have greatly improved sensitivity. The longer wavelengths enable JWST to look further back in time to see the first galaxies that formed in the early universe, and to peer inside dust clouds where stars and planetary systems are forming today.

The James Webb Space Telescope will be a giant leap forward in our quest to understand the Universe and our origins. JWST will examine every phase of cosmic history: from the first luminous glows after the Big Bang to the formation of galaxies, stars, and planets to the evolution of our own solar system. The science goals for the Webb can be grouped into four themes:

- **The End of the Dark Ages: First Light and Reionization** - JWST will be a powerful time machine with infrared vision that will peer back over 13.5 billion years to see the first stars and galaxies forming out of the darkness of the early universe.
- **Assembly of Galaxies** - JWST's unprecedented infrared sensitivity will help astronomers to compare the faintest, earliest galaxies to today's grand spirals and ellipticals,

helping us to understand how galaxies assemble over billions of years.

- **The Birth of Stars and Protoplanetary Systems** - JWST will be able to see right through and into massive clouds of dust that are opaque to visible-light observatories like Hubble, where stars and planetary systems are being born.
- **Planetary Systems and the Origins of Life** - JWST will tell us more about the atmospheres of extrasolar planets, and perhaps even find the building blocks of life elsewhere in the universe. In addition to other planetary systems, JWST will also study objects within our own Solar System.

Courtesy NASA Web-site - **https://www.nasa.gov/mission_pages/webb/science/index.html**

Now let's put our new Mission Success idea into action. As a member of the new NGST (Now JWST) Program, wearing my Systems Engineering hat, I had the task of reviewing several mission critical tasks. First, I was assigned to review and update, using our new MS process, the Systems Engineering Management Plan (SEMP) for short. I was appalled of the poor job someone had done with a first draft. Using International Council on Systems Engineering (INCOSE) Book of Knowledge (SEBook) references, my NASA Standard Assurance Management Office practices, and 40 years in the business, I literally tore it apart. Starting with a trust, but verify mindset, I added in all new part, sub-system, systems testing, validation and Risk Mitigation sections to traditional PDR (Preliminary Design Review), Critical Design Review (CRD), Flight Readiness Review (FRR) engineering reviews. My team members really beefed up the contractor requirements with Assurance Management criteria and that insured Space level Quality Control criteria was followed by NASA and contractors throughout the design, fabrication and build processes.

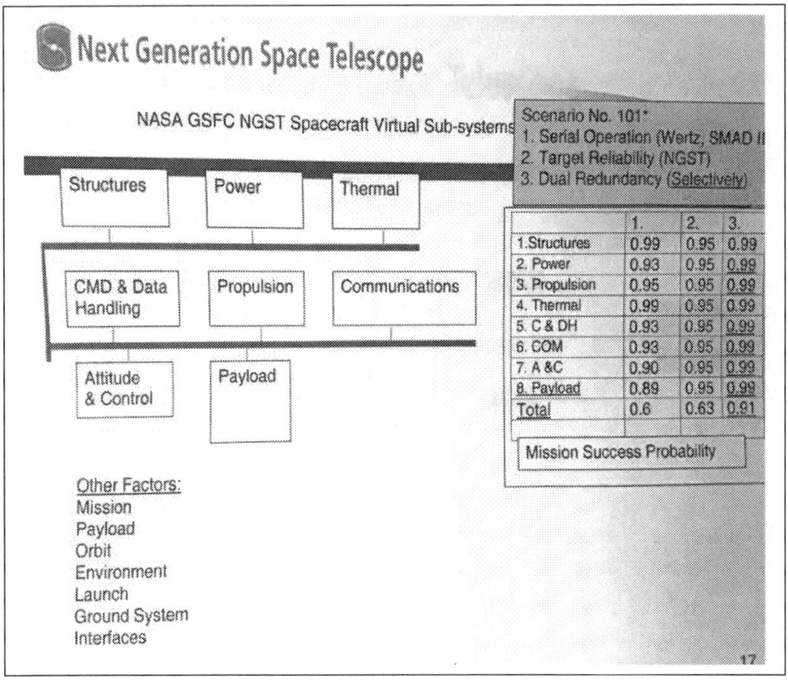

Exhibit MS – 50: JWST Spacecraft Sub-system MS Probability.

Exhibit MS – 50 shows the 8 JWST major sub-systems and targeted reliability levels for each of them. For Mission Success to be achieved, a recommended three levels deep in design is recommended.

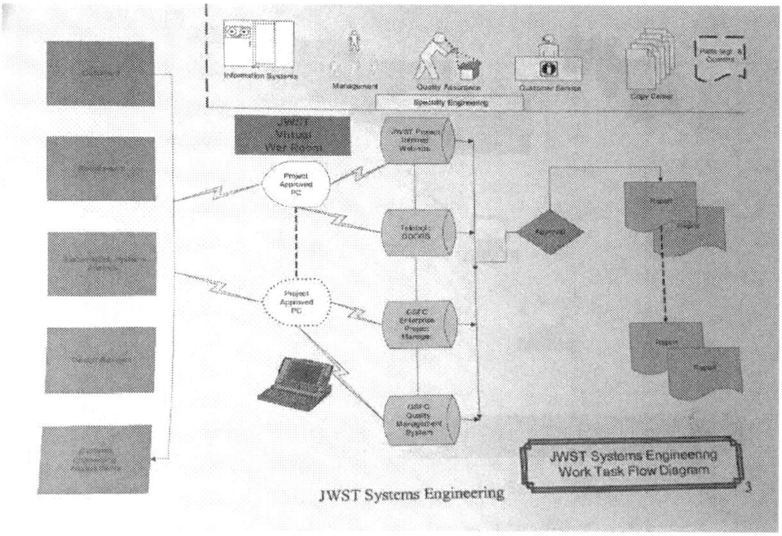

Exhibit MS – 51: JWST Systems Engineering Work Task Flow Diagram

Exhibit MS – 51 above lays out in flow diagram flow the Work Task required by Systems Engineering and the functional group relationships of other team members.

This leads us to my next work task was the Software Engineering Management Plan (SEMP).

I started in the computer business in the early 1970s and even taught programming and Management Information Systems at the graduate school level. And, in my first management position at NASA, both hardware and software engineering personnel reported to me. I even put a hardware graduate EE over the Software folks to assure checks and balances existed with the software engineers. My EE also had software background and knew enough to be dangerous. It worked out just fine and eventually Software Assurance grew to a much bigger group and was spun-off to a separate reporting software organization.

Today, software is as big a group as it's hardware counterpart and with software driven design more of a Systems Engineering type organization exists.

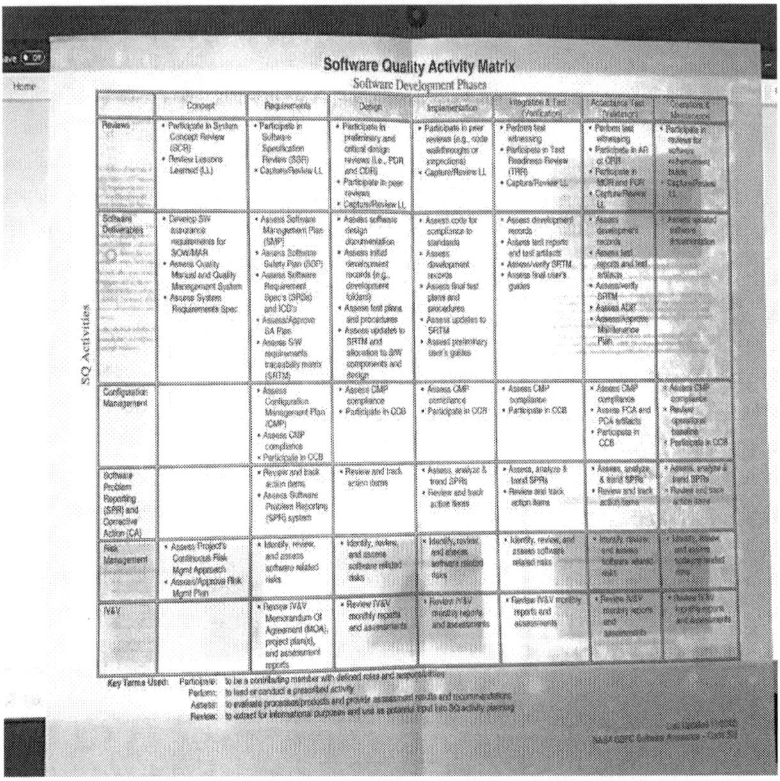

Exhibit MS – 52: Software Quality Activity Matrix

Using the above Matrix (Exhibit MS-52) that covered Reviews, Deliverables, Configuration Management, Problem Reporting, Risk Management & Independent Verification and Validation (IVV) I was given a roadmap to evaluate requirements documents for compliance to requirements.

I was not happy with what I saw with the Software Development Plans. Again, a boiler plate strawman procedure was attempted. On my first pass, I wrote up a six-page list of shortcomings. With such a large-scale, multi-billion-dollar project as NGST (now JWST) was going to be, I felt a more comprehensive plan needed to be written.

As Boeing is taking such a hit today over the 727MAX software supposedly being sub-contracted to Pakistan for $7 dollars per hour…I hope that is not true! I had the same feeling when faced with a new software supplier from New England that was a new organization. After checking their finances, I found out that the company was in deep debt and could not understand why they were being used. In fact, several military contractors on the Beltway were also using this firm. I discussed my concern with top management and later found out that the new company was bought out by a Blue-Chip computer firm. So much for that venture.

I went back to the Software requirements and Software Management Plan and again embedded MS, QMS, Risk Management (PRA/FTA), use of Beta Site, SEI & CMM approaches, security and virus protection and verification/validation practices along with testing and acceptance criteria. With hardware, it is easy to find in workmanship errors and part failures by inspection and testing. With software not so easy, why do you think so many versions exist from 1.0 to 5, 6, 7.0, etc. We used to call it "de-bugging" an older generation computer term! Ask Boeing about that one!

One excellent resource I bought was a book called ***"Software Project Survival Guide"*** written by Steve McConnell and published by Microsoft Press (See: construx.com). His Chapter 9 covers a myriad of Quality checks to assure specification compliance.

The next series of charts will show a more detailed flow and roadmap of our actions.

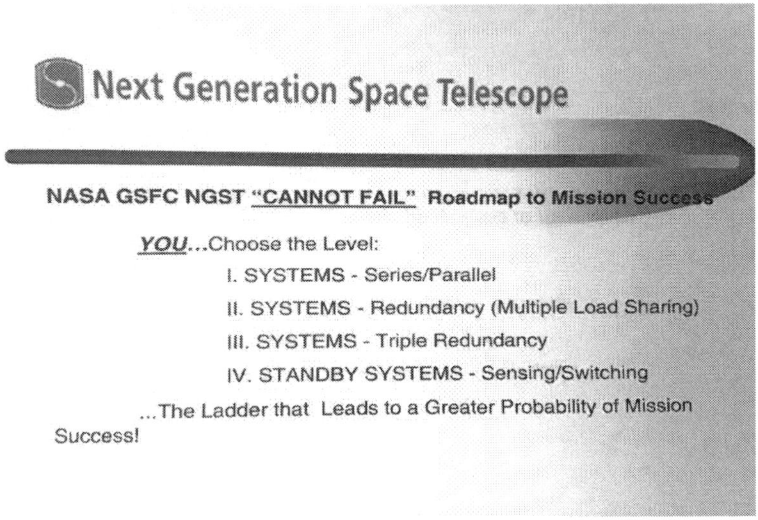

Exhibit MS – 53: Roadmap to Mission Success

As illustrated in Exhibit MS – 53, there are four levels to be implemented:

The first three deal with systems design and recommends parallel paths to design including built-in redundancy, double or triple based on Risk studies and the use of self-healing circuitry and switching to reduce failures and increase probability for mission success.

So, let's review our original presentation for what we did and how we did them, and why!

With almost a $10,000,000,000, yes, a $10 Billion-dollar Program running for almost 20 years (original start date 1996 and budget of ~$900M, re-designed 2005) our contributions for embedding Mission Success factors we hope, pays off big time. Launch is now scheduled for March 20, 2021 and we will see the next HST generation, James Webb Space Telescope, launched into deep space for new cosmetology and astronomy investigations. Mission Duration is 5 years with a 10-year goal.

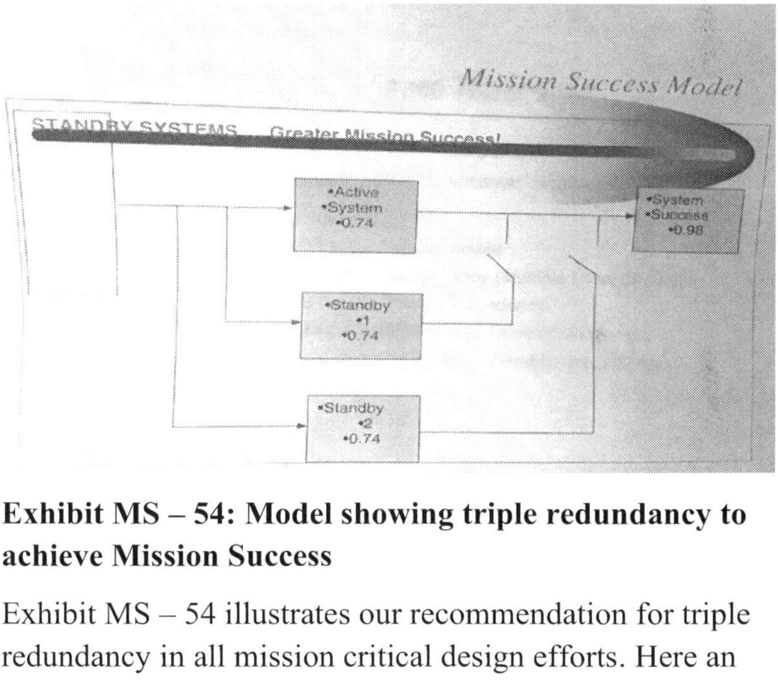

Exhibit MS – 54: Model showing triple redundancy to achieve Mission Success

Exhibit MS – 54 illustrates our recommendation for triple redundancy in all mission critical design efforts. Here an active system failure automatically defaults to Standby #1 system. In case of failure there again default goes to Standby #2 that assures seamless operation. Got that Boeing? You had it for years, what happened? Perhaps that cheaper, better, faster mantra got in the way like it did for years at NASA! Stick to the past best practices in design and manufacturing, it works every time!

At the time, I was assigned to the JWST Program, I was given the opportunity to review and incorporate many of the ideas contained in this book in the latest NASA Systems Engineering Handbook. The final version updated is on-line at Amazon.com and there is even a lower price Kindle version available. I downloaded the latest version and highly recommend it for your library and personal use.

So, all the requirements, practices, training, inspection, testing, reviewing we have covered are now embedded into the new space telescope. Provided they will continue, perhaps a lifetime as specified will be accomplished and exceeded, just like its predecessor the Hubble Space Telescope. For more detailed spacecraft information, see NASA Educators Website on JWST:
https://jwst.nasa.gov/content/forEducators/formal.html

In addition to the SEMP and SMP, I had the opportunity to take a shot at creating a WEB site design for the JWST SE management to use to oversee work progress (See Exhibit MS – 55).

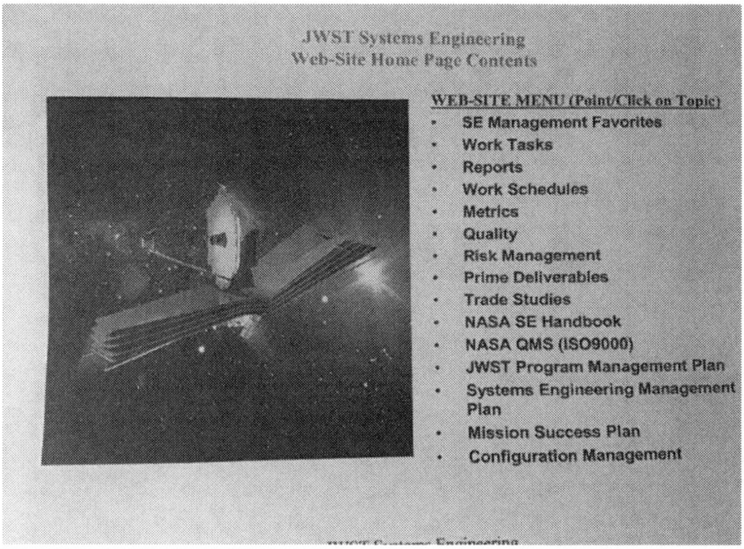

Exhibit MS – 55: SE Web-Site Home Page Contents for JWST Program

We finish this chapter by reviewing NASA GSFC missions my teams at USA contractor sites and KSC/VAFB and I supported during my NASA 18-year timeframe from 1989 to 2008. Many like ACE are still operational as of 2019. All have had measures of mission success requirements associated with the design, build and launch phases of their life cycles.

One had anomalies with Commercial Off the Shelf (COTS), unqualified parts, that caused failure and for further information see the Failure Review Board Report on ICESAT.

Another like TDRSS 2, was lost in the Challenger Disaster in 1986. I recommend JPL's writeup on Challenger for further study (https://er.jsc.nasa.gov/seh/explode.html). I was in Florida when Challenger was launched. The night before (January 27,1986) was one of the coldest nights I ever experienced there. I recall even wearing my winter northern bought leather bomber jacket to keep me warm and scrapping my car windshield to clear off the ice that was forming.

For Columbia, a second shuttle tragic event, the Columbia Accident Investigation Report (CAIB) provides detailed information on the Investigation. (https://www.nasa.gov/columbia/home/CAIB_Vol1.html).

For the most part, NASA GSFC has a fine track record when it comes to overseeing and building Satellites. And with their surveyed and qualified contractor base of the world's best aerospace suppliers, they continue to supply GPS, Weather, Tracking and Communications products second to none when it comes to Quality and Reliability.

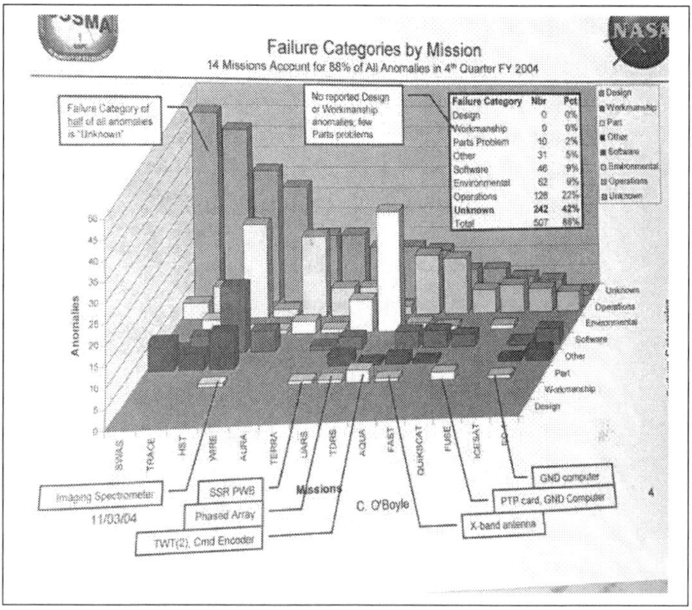

Exhibit MS – 56: Spacecraft Mission Anomaly Analysis (One Quarter FY'04)

Typical computer driven analysis so nicely displayed in a 3D Graphic in Exhibit MS – 56 shows little design and workmanship problems, however, active parts do have a history of failure.

Whether due to Space environment conditions, early life or simply wear out. Therefore, redundancy in design is so important for long life missions.

A while back, I researched, at the time, contemporary high technology Systems failures and offer my list of drivers for your review. Operator error was a big contributor and today it still is as in auto accidents where drivers are distracted by the many devices at their disposal. Even 911 could be one of our biggest ever with lack of system barriers offered a big opportunity for terrorists to overcome easy to penetrate processes.

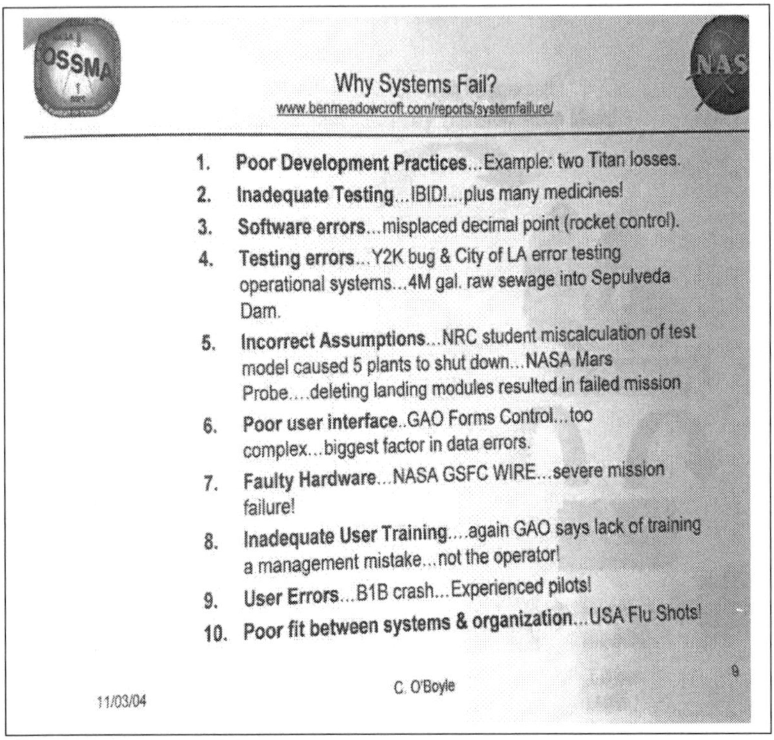

Exhibit MS - 57: Why Systems Fail? (http://www.benmeadowcroft.com/reports/systemfailure/)

I credit Ben Meadowcroft, MESE, MBA, here (Exhibit MS – 57) as I used his early model to site a few contemporary problems that were occurring in the world in those days. .

Today, his website covers the software area and again he does a swell job of listing problems in that arena as he did in his earlier systems studies.

Exhibit MS - 58: NASA GSFC Missions My Teams Supported.

ACE	Delta II – 8/25/97	JHU/APL
AQUA	" - 5/4/2002	Northrup Grumman
AURA	" - 7/15/2004	Northrup Grumman
GOES 8	Atlas Missile – 4/13/94	Space Systems Loral
GOES 9	" - 5/23/95	"
GOES 10	" - 4/25/97	"
GOES 11	" - 5/3/2000	"
GOES 12	" - 7/23/2001	"
GRO	Atlantis STS-37 – 4/5/91	TRW
HST SM 1	Endeavor STS-61 - 2/2/93	Lockheed Martin/GSFC
HST SM 2	Discovery STS-82 - 2/11/97	GSFC
HST SM 3A	Discovery STS-103 - 12/19/99	GSFC
HST SM 3B	Columbia STS-109 – 3/1/03	GFSC
HST SM 4	Atlantis STS-125 – 5/11/09	GSFC

ICESAT	Delta VAFB – 1/13/03	See FRB Report (COTS)
JWST	To be launched 3/30/2021	Ariane 5 ESA Fr. Guiana
LandSat 7	Delta VAFB – 4/15/99	Lockheed Martin
LandSat 8	Atlas V VAFB 2/11/2013	OSC Corp.
LRO	Atlas V KSC 6/18/2009	NASA GSFC
NOAA N	Delta VAFB 5/20/2005	Lockheed Martin
ROSAT	Delta KSC 6/1/90	GR, UK, USA mission
SOHO	Atlas II KSC 12/2/95	ESA/NASA Team
SORCE	Pegasus KSC 1/25/2003	OSC Corp.
Stereo	Delta II KSC 10/26/06	Johns Hopkins Labs APL
Swift	Delta KSC 11/20/2004	Penn State Flight Op Team
TDRSS (# 3 -7 Satellites) *	All Shuttle Launches KSC	TRW Redondo Beach, CA
TERRA	Atlas II VAFB 12/18/99	Lockheed Martin
THEMIS	Delta KSC 2/17/2007	OSC ATK
TIRED	Delta VAFB 12/7/2001	APL Johns Hopkins
TRACE	Pegasus XL KSC 4/1/98	OSC/ Lockheed Martin

WIND	Delta II KSC 11/1/94	Martin Marietta
NASA/USA/Japan Launch	Tanegashima, Japan	
TOMS Meteor 3	Plesetsk, Russia 12/94	Cyclone Booster
*TDRSS # 2 lost on launch	Challenger 1/26/1986	STS-51L

As one can see by column three on the Exhibit MS - 52, many Aerospace contractors are involved in NASA GSFC missions. At one time, I had 15 locations across the USA as NASA In-Plant Representatives, who acted as NASA's eyes and ears on spacecraft work located there. At KSC, we also had our GSFC Representatives located there overseeing all prelaunch activities occurring right up to launch day. At VAFB, Project Representatives were TDY to the Lompoc, CA facility. International launch missions had my teams accompany the spacecraft shipment to the launch sites, some of them in Japan and Russia. I am pleased to report every one of these missions were highly successful thanks to their oversight.

Ed. Update: I reviewed the present NASA requirements and are pleased to see almost all of what we covered in this book on mission success was incorporated in high level documents and requirements used today in our, still best on the planet, space satellite programs.

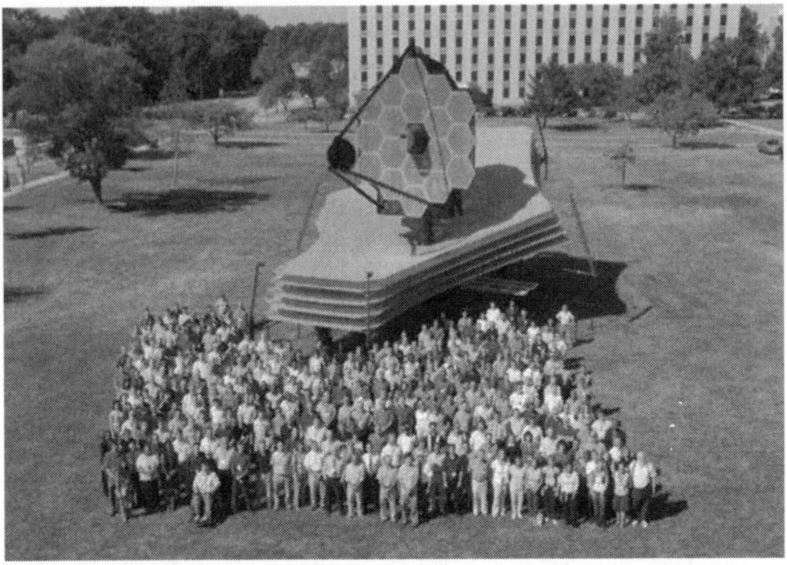

Exhibit MS – 59: NASA GSFC James Webb Space Telescope Team (me too!) & Mock-up!

Exhibit MS – 59 shows original JWST Project Team assembled in front lawn of NASA GSFC Headquarters with Mock-up of Spacecraft in background.

References:

TRW Space Log – TRW Space & Electronics Group, Redondo Beach, CA 90278 1995

Space Systems Failures - Harland & Lorenz, Praxis Publishing Co. Chickester, UK 2005

The Challenger Launch Decision - Diane Vaughan, Chicago University Press 1996

Failure is Not an Option – Gene Krantz, Mission Control Director, Simon & Shuster NY 2000

50 Years of Mission Assurance *Presentation 2007 NASA GSFC Contractor's Workshop*

NASA Web-Site: nasasupplierconference2007RevA Chuck OBoyle/Jim Suraci Presentation

Software Project Survival Guide by Steve McConnell Microsoft Press 1998

Quality is Free: The Art of Making Quality Certain by Phil Crosby 1979

My 50 Years in High Technology by Charles V. O'Boyle (to be published soon!)

No BS…Your MBA Primer by Charles V. O'Boyle (to be published soon!)

Out of the Crises by Dr. Edward Deming MIT Press 1982

Space Vehicle Systems & Engineering Handbook by Aerospace Corporation 2006

Rules for Design, Development, Verification & Operation of Flight Systems by NASA GSFC-STD-1000 December 10, 2004

The Six-Sigma Way by Pande, Newman & Cavanaugh McGraw Hill, NY 2000

Development of a Space Vehicle Test & Evaluation Handbook **by** Peterson, Hanifen & Tosney Aerospace Corporation and Lt. Col. Susan E. Hastings, NRO AIAA 2007-1630

Hazard & Barrier Analysis Guidance Document (EH 33) by the DOE Office of Operating Experience Analysis & Feedback Nov 1996

Barrier Analysis Guidance - web-site https://sbms.bnl.gov/SBMSearch/subjarea/39/2m04e011.doc

ISO Information:

http://ts.nist.gov/Standards/Global/ir4721.cfm

NASA Websites:

Headquarters: **www.nasa.gov**

GSFC: **www.nasa/goddard.gov**

KSC: www.nasa/KSC.gov

The Author's Background

Chuck O'Boyle has 20 years U.S. Government service (18 at NASA and 2 in the US Army Ordnance Guided Missile School in Huntsville, Al).

Prior to his NASA experience, Chuck was Quality Assurance Director at Kollsman Instruments, Quality Engineering Manager at Martin Marietta, Operations Manager at Teledyne and QA & Reliability Manager at Sanders Associates (ex-Lockheed Martin, now BEA). Early in his career, he worked as a Sr. Engineer for the Hewlett-Packard Company specializing in microwave instrumentation.

Chuck worked on the Pershing, Patriot, Titan, B1-B, M60 Tank Fire Control System, Tomahawk, Harpoon, USAF F-111 & US NAVY ALQ-126 Electronic Countermeasures Systems & Nike Ajax/ Hercules Guided Missile Systems. At NASA, he worked close to 100 successful missions including HST SM # 1 - 3A, JWST, GRO, XTE, EOS AQUA, AURA & TERRA, TRMM, LandSat 7 & International Space Station Fluids & Combustion Facility receiving numerous NASA Achievement Awards for his engineering/management contributions and efforts. The HST QE Team he managed won 36 NASA Medallions for their work on the highly successful Servicing Mission No. 1 Repair Mission.

For the past several years, as a ManTech SRS Lead Engineer, he supported GSFC Mission Operations for the Office of Mission Assurance and Systems Engineering organizations covering all GSFC on-orbit mission performance and the new James Webb Space Telescope (JWST) Program.

Chuck taught part-time at the Graduate School level in the Boston area for 10 years and is an Honors Graduate of Northeastern University AS & BS and Babson College MBA, with Distinction. He is also a graduate of both Martin Marietta and Unisys Advanced Management Programs. And he is a member of Sigma Epsilon Rho National Honor Society

Now retired, but still is writing his books. He spends his summers in PA and Maine and winters on the Gulf of Mexico in Florida, USA.

Made in the USA
Columbia, SC
11 January 2023

10130221R00083